ICONIC
MOMENTS
IN BROADCAST HISTORY

ICONIC MOMENTS

IN BROADCAST HISTORY

LIVE VIA SATELLITE

ROBERT M. PATTERSON

ISBN (paperback): 979-8-9943455-0-4
ISBN (hardcover): 979-8-9943455-1-1
ISBN (ebook): 979-8-9943455-2-8

Book design and production by www.AuthorSuccess.com
The front cover control room image was rendered in part by AI.
Cover photo of two astronauts retrieving a satellite from space: Photo courtesy of NASA.

ACKNOWLEDGEMENTS

To Marlin Fitzwater,
Of all the remarkable people I have met during my career, I could not be prouder than to have Marlin Fitzwater write the foreword for this book. Unbeknownst to us, our career paths had already intersected during the Reagan and Bush presidencies, and years later, we would come together as fellow trustees at Franklin Pierce University. A genuine bond grew as I listened to his extraordinary stories, observed his graciousness in helping others, and witnessed the deep respect and admiration he earned from journalists and peers alike. He is a master of humility and self-assurance, and I am truly honored to call him my friend.

To My Family,
I wish to express my profound gratitude to my son, Rob, whose unwavering encouragement, invaluable contributions, and insightful perspective have been instrumental in bringing this book to fruition. His guidance, both intellectual and personal, has been the cornerstone of this project. Without his support, this work would not have been possible, and for that, I am forever grateful.

To my loving wife, Nancy: You have been my steadfast companion through every triumph and challenge, offering understanding, strength, and support that have carried me forward. Your devotion has been both my inspiration and the foundation of my journey.

Together, you have granted me the most cherished gift of all: the fulfillment of a lifelong dream and enduring happiness.

With all my love,
Bob

HOW SATELLITES IGNITED A GLOBAL AWAKENING

"Satellites are the heartbeat of global communications;
without them, humanity's ability to connect and
collaborate would cease to exist."

Robert M. Patterson

CONTENTS

FOREWORD

On June 23, 1998, the President of Franklin Pierce University invited his board of trustees, students, and faculty to a dinner at the Ronald Reagan Building in Washington, D.C. The guest of honor was President George H.W. Bush, and among the hundreds of guests was another president: Robert M. Patterson. He was perhaps the most humble of all the guests and played a significant role in shaping the modern communications and broadcasting industries.

And now, in 2025, he has published a shining new book that tells his personal story and the history of satellite communications in America. From the first commercial satellite, the Early Bird, that soared into orbit nearly sixty years ago, Bob Patterson has been a pioneer in satellite communications, and virtually every home in America has benefited from programming supported by satellite.

Robert Patterson has supported the Marlin Fitzwater Center for Communication at Franklin Pierce University, developed the Patterson Television Studio, and contributed to numerous enhancements for education and satellite communications.

He is also an excellent writer, as this book demonstrates.

Marlin Fitzwater
Fitzwater Center for Communication 2025

INTRODUCTION

This day, Saturday, August 9, 2025, I sit quietly, marking the fiftieth anniversary of an event that profoundly shaped my life. On this day, I arranged and conceived the first-ever live multi-carrier telecast via domestic satellite, an achievement that transformed the way sports, news, and entertainment are broadcast in the United States.

From my desk, with a view of the Pacific Ocean, the endless waves ebb and flow like the passage of time, carrying me back to another era, when I was growing up in Queens, New York. Summer days were spent playing stickball in the schoolyard, nights chasing fireflies, and dreams that stretched far beyond the city skyline.

In those days, I imagined myself as a space explorer like John Glenn, charting courses through distant galaxies. Other times, I was Mickey Mantle at the plate, eyes locked on an invisible fastball, swinging for the fences with everything I had.

Fate, however, had other plans. I soon realized that I didn't have the "right stuff" to be an astronaut, nor the talent to become a sports superstar. Yet destiny has a way of charting its own course. Over time, I discovered a way to merge my passions for space exploration and sports into a single career: satellite broadcasting.

I've learned that the road to any milestone is rarely a straight line. It twists and turns through the unexpected, carrying us forward into moments of doubt and discovery, trials and triumphs. Looking back now, I see clearly that it was the road itself that defined my journey.

On a crisp October evening in the late '50s, the world shifted forever. A small, shining sphere, with four slender antennas, glided across the night sky, sending out eerie, rhythmic beeps that echoed around the globe. It was October 4, 1957, the night the Soviet Union launched *Sputnik*, the world's first satellite, and the birth of the Space Age.

At the time, I had no way of knowing that those distant signals would one day shape the path of my life: a career in satellite communications spanning more than five decades. I would find myself with a front-row seat to some of the most iconic moments of modern history, connecting the world to events that would define generations.

Iconic Moments in Broadcast History: Live Via Satellite takes you on a thrilling journey through the evolution of satellite technology, told through the eyes of someone who not only witnessed its rise but helped shape the modern communications and broadcasting industries.

Together, we will travel through the first eighty years of satellite history, from Sir Arthur C. Clarke's visionary concept to the groundbreaking technology of today that touches nearly every aspect of our daily lives. Along the way, you'll step inside the events that changed the world, not just as a viewer, but as an insider.

This odyssey is filled with untold stories of satellites spinning out of control, high-stakes confrontations behind closed doors, personal triumphs and crushing setbacks, and the frantic chaos that erupts in control rooms when live global broadcasts vanish from the airwaves without warning.

I share this story as a tribute to the unwavering commitment of the men and women who built, maintained, and advanced the satellite communications industry. These unsung heroes made it possible for the world to witness history as it happened, whether in moments of triumph, heartbreak, or urgency, ensuring that live sports, breaking news, and vital information reached our homes with flawless precision.

Their work was often invisible, yet its impact was immeasurable. Behind every broadcast was a quiet army of professionals working in the shadows, driven by a shared mission to keep the signal alive. They asked for no recognition, content in the knowledge that hundreds of millions of viewers could share in the defining moments of our day; moments that might have otherwise been lost to distance and time.

This is their story, and mine. *Iconic Moments in Broadcast History: Live Via Satellite* is both a passage through the extraordinary evolution of technology and a tribute to the people who made it possible: those who brought the world closer together, one signal at a time.

CHAPTER 1

A SATELLITE ODYSSEY

THE OLYMPICS

It was July 28, 1984. The sun blazed high above the Los Angeles Coliseum as the world gathered for the Opening Ceremonies of the Games of the XXIII Olympiad. Inside the vast stadium, history seemed to hang in the air. It had been fifty-two long years since Los Angeles last hosted the Olympic Games, and this day was destined to be etched into the annals of history, an unforgettable moment that would resonate for generations to come. The stadium pulsed with energy, packed with cheering fans, Hollywood stars, and Olympic legends returning to the stage that had once immortalized them.

The world was watching. This was more than a ceremony; it was a beacon of hope and unity in a world divided by Cold War tensions. But here, in this stadium, politics faded into the background, and the spirit of international friendship took center stage.

The world held its breath as athletes poured into the storied Los Angeles Memorial Coliseum, a river of color and emotion flowing beneath a golden California sky. Their faces beamed with triumph and their flags unfurled in a dazzling blur of national pride. Cameras panned and zoomed, capturing the heartbeat of history and transmitting it live to a global audience. The Opening Ceremony burst to life with soaring balloons, fluttering doves, and a wave of motion that transformed the field into a living work of art.

For those watching the Parade of Nations at home, it was perfection in motion. The world was spellbound as the Olympic spectacle unfolded in breathtaking grandeur. Every frame, every bit of fanfare, and every flicker of emotion was broadcast seamlessly, drawing viewers across every corner of the globe into the magic of the Games of the XXIII Olympiad.

Then, without warning, the unthinkable happened.

Television screens suddenly went black as millions around the world lost the signal. Confusion rippled across the globe as an eerie, inexplicable silence replaced the joyous spectacle.

Now, imagine yourself in the control room, the nerve center of a global broadcast, charged with delivering the 1984 Summer Olympics to more than a billion viewers. Then, in an instant, the screen goes dark.

This was the reality I faced as president of a satellite communications company, entrusted with the monumental task of delivering the games to audiences across Asia, Oceania, South America, and Europe. Everything I had worked for—my career, the Olympic Games, and the broadcast—hung by a fragile thread.

To better understand that heart-stopping moment when the Olympic broadcast vanished from the airwaves, let's rewind to early 1984, when I was crisscrossing continents, negotiating with foreign broadcasters, and meticulously assembling the global transmission network that would deliver the Olympic Games to the world.

A Whirlwind of Negotiations

While ABC held exclusive broadcast rights in the United States, I was responsible for navigating the intricate task of securing transmission rights for international broadcasters worldwide.

This was no ordinary assignment; it was a high-stakes balancing act played out on a global stage. Securing transmission rights in more than fifty countries felt like scaling Mount Everest in a suit and tie. Every conversation required finesse, timing, and the perfect blend of persuasion and diplomacy.

These weren't just meetings; they were mental marathons, testing my stamina, willpower, and, at times, my sanity. Travel became a blur of airports, with days slipping away in a jet-fueled haze and nights offering no relief, just room service, jet lag, and the shriek of an early alarm. I wasn't living, I was surviving—driven by caffeine, adrenaline, and an unwavering mission: to connect the world to the Olympic Games.

The first major hurdle was the European Broadcasting Union (EBU), a formidable coalition of thirty-five nations, each with its own agenda, regulatory maze, and carefully guarded expectations. Negotiating with the EBU was far more than a matter of securing satellite access; it was a high-stakes diplomatic contest veiled in strategic maneuvering. This was a battlefield where politics clashed with production, where a single contract clause could unravel delicate alliances and one misstep could undo months of painstaking effort.

The negotiations reached their apex in Geneva, Switzerland, home of the European Broadcasting Union. After days of intense bargaining, tactical maneuvers, and guarded concessions, we finally sealed the deal.

To mark the milestone, we gathered for a lakeside luncheon along the tranquil shores of Lake Geneva, a rare moment of pause in an otherwise unrelenting marathon. Nearby, the Jet d'Eau, the city's iconic fountain, soared 160 feet into the air, its spray catching the sunlight. For a fleeting moment, I allowed myself a quiet sense of accomplishment. However, the mission was far from over; broadcasters still awaited, signals needed alignment, and continents had yet to be connected.

Asia loomed on the horizon: a complex tapestry of tradition and bureaucracy, where what seemed like simple negotiations often transformed into elaborate rituals and painfully slow approvals. Here, patience was more than a virtue; it was a calculated strategy. In Tokyo, respect was won through quiet discipline and strict adherence to hierarchy; in Taipei, trust was forged over shared meals and long, late-night sessions, fine-tuning every technical detail.

Oceania moved to its own steady beat. In Sydney and Melbourne, broadcasters were pragmatic and passionately patriotic, treating the

Olympics as a matter of national pride. For them, the Games were sacred; a moment when every satellite path and backup wasn't merely checked but fiercely guarded. Clarity was expected, honesty was their way of life, and above all, unbreakable trust was the foundation of every broadcast.

South America moved to a different beat. Brazil was vibrant, yet its infrastructure was uneven, and approvals were often trapped in the molasses of bureaucratic red tape. Progress happened not in boardrooms but over late-night dinners and informal conversations, where relationships mattered as much as signatures.

By the time I finalized agreements with Brazil, Australia, Taiwan, Japan, and Europe, a quiet sense of accomplishment began to take hold. I had successfully navigated the technical, cultural, and political intricacies of broadcasting with some of the world's most influential nations.

Just as the weight began to lift, I was reminded that the most formidable challenge still lay ahead: the Soviet Union.

In 1984, the Cold War wasn't merely a geopolitical rivalry; it was a high-stakes chess match played on the razor's edge of a knife. Every negotiation between the United States and the USSR carried the weight of the world, both literally and symbolically. The atmosphere was thick with suspicion; every word was parsed, every gesture scrutinized. These talks weren't about resolution, but posturing. Each side was determined not to blink first.

I found myself seated across from the Soviet Director of the Ministry of Communications, an imposing figure with eyes like ice and an expression that revealed nothing. He was more than a negotiator; he was a gatekeeper for the Kremlin, tasked with safeguarding its interests and extracting every possible advantage. For him, this meeting wasn't about Olympic spirit; it was a battle for control, perception, and propaganda.

The air in the room was heavy with unspoken tension, as if the walls themselves were listening. We sat deep inside a drab, colorless conference room, flanked by translators whose words were measured and meticulous. They were charged with carefully wording each response by the director, who seemingly knew little English. Behind them, protocol officers stood stiff and watchful, like sentries guarding a ceasefire that could collapse at any moment.

Then, something shifted.

An assistant entered quietly, placing a frosted bottle of vodka on the table. There were no words, no smiles, just the weight of ritual. The director poured slowly, the silence louder than words, then slid a shot across the table, initiating a test I hadn't anticipated.

We raised our glasses, not in celebration, but in acknowledgment. Then, a subtle but unmistakable change: the faintest curl at the corner of his mouth. A smile. In that moment, something softened, the tensions of treaties and talking points giving way to a genuine human bond.

As the vodka flowed, so did conversation. He revealed that he represented more than just Moscow. He spoke for the Organization of International Radio and Television (OIRT), a sprawling coalition of Eastern Bloc broadcasters, spanning East Berlin to Havana, Hanoi to Prague. With each glass, his English improved. By the fourth shot, the translators had become obsolete.

Each round drew us closer. Tensions ebbed. Ideas aligned. Hours passed, and the table gradually filled with empty glasses. Eventually, we had more than a handshake—we had an understanding. Not just a document, but a commitment. The Soviet Union would carry the Olympic Games, and so would its allies.

When the final glass hit the table, so did the signed contract. More than a business deal, this was a diplomatic breakthrough, sealed not by protocol, but by trust, one shot of vodka at a time—something I would later characterize as *vodka diplomacy*.

The Task Ahead

The signatures were dry, but there was no time to celebrate.

Securing the transmission rights was only the beginning. The real challenge lay in delivering the games live to a global audience. Much more than a technical objective, it was a commitment to unite the world in a shared, real-time experience.

All systems go, frequencies aligned, contingencies in place, signal paths verified. The satellites were poised above, and the world waited below. Then, May 8, 1984, arrived like a storm: sudden, uninvited, and unwelcome.

A cold, stark message from Moscow tore through months of fragile diplomacy like a thunderbolt, shattering alliances, unraveling painstaking negotiations, and obliterating the trust forged over countless shared toasts. In an instant, everything we had built so carefully had vanished.

The Soviet Union announced that it would boycott the Los Angeles Summer Olympics, claiming concerns over the safety of its athletes and accusing the United States of fueling anti-Soviet hysteria. But behind the carefully worded statement, the true motive was unmistakable: retaliation. Four years earlier, the U.S. had led a boycott of the 1980 Moscow Games in protest of the Soviet invasion of Afghanistan.

The fallout was immediate and brutal. The games continued, but behind the spectacle and celebration, a heavy shadow lingered.

While athletes trained in pursuit of glory, behind-the-scenes political maneuvering ultimately determined who stepped onto the field.

THE BIG DAY ARRIVES: July 28, 1984

As Los Angeles unveiled the Games of the XXIII Olympiad, the world watched history unfold in real time.

Then—everything stopped.

Inside the control room, elation evaporated. Red lights blinked like alarms on a sinking ship. **"NO SIGNAL"** flashed in bold, screaming letters. Heads snapped toward the monitors. Panic rippled through the room.

The Parade of Nations, the emotional heart of the Opening Ceremony, had just begun. Flags were waving. Athletes were marching. But no one beyond the Coliseum could see a thing.

"Where's the feed?" someone called, already fearing the answer.

The satellite transponders had gone dark. This wasn't a minor glitch;

it was a full-blown broadcast crisis. The most-watched live event on Earth had suddenly disappeared from the airwaves.

Technicians sprang into action. Fingers flew across keyboards. Backup systems engaged. Signals were scrambled, rerouted, and redirected. However, the screens remained black.

Seconds dragged into minutes. Around the globe, viewers sat in confused silence, staring at their blank screens.

Inside the operations center, it was an all-out emergency.

Every second was unbearable. For the global audience, it was little more than a mysterious hiccup, but for us, it was a nightmare playing out in real time.

We had contingencies and backups, but none of them were restoring the feed.

Faces tensed, eyes darted across monitors, desperate for the faintest flicker of life. Overhead, the clock ticked on, relentless and unforgiving.

Failure was not an option. The world was watching . . . except now, it wasn't.

Then, the control room door flew open. An engineer from the European Broadcasting Union rushed in, breathless but beaming. "We're back up and running!"

A collective exhale swept through the room as tension gave way to relief. The heartbeat of the broadcast surged back. Monitors flared to life, colors flooded the screens, and the sound roared back to life.

High above Earth, a constellation of satellites danced in synchronized orbit. Invisible beams of data surged from ground uplinks to satellites and back down to Earth, crisscrossing oceans and borders in microseconds. The world was reconnected.

On-screen, the Parade of Nations, once silenced, now streamed in full glory, reviving the spirit, the spectacle, and wonder of the moment.

As the final athlete stepped in place, a hush fell over the Coliseum. The cameras cut to a pre-recorded message from President Ronald Reagan. His voice, resolute and proud, echoed across the globe via satellite: "I declare open the Olympic Games of Los Angeles, celebrating the XXIII Olympiad of the modern era."

The Olympic flag rose. The anthem played, and the Coliseum held its breath. Then came the final act—the emotional crescendo.

Olympic champion Rafer Johnson emerged with quiet majesty, the torch held high, its flame dancing against the twilight. His eyes fixed on the summit, Johnson began his ascent, each step purposeful, a solemn march toward history. At the peak, with unwavering grace, he touched fire to the cauldron, unleashing a radiant plume that soared into the sky, igniting not only the Olympic flame, but the very spirit of the Games.

The stadium erupted in thunderous celebration, while across the globe, billions watched from their living rooms, spellbound.

SOLVING THE MYSTERY

Behind the scenes, the search for answers was underway, a high-stakes race against time to identify the problem and find the culprit.

Was it a power surge? A transponder failure? Human error? Or something more sinister: sabotage?

Theories flew as fast as the crisis had hit. While the world watched athletes parade beneath fluttering flags, we pored over signal logs, inspected power grids, and traced every cable like detectives on a case.

We braced for a complex answer, some deeply embedded flaw in the system.

What we found instead was so absurd, so unexpected, it felt like the punchline to a cruel joke.

The culprit? A $4.99 metallic party balloon. Symbols of celebration, shimmering, helium-filled, tethered only to joy, had nearly taken down the most-watched television event on Earth.

During the Opening Ceremony, in a moment choreographed for maximum spectacle, thousands of metallic balloons were released into the Los Angeles sky. It was meant to dazzle: a soaring tribute to unity and festivity. Nature, however, had other plans. Some of the balloons drifted off course, colliding with high-voltage power lines, triggering a spectacular arc of electricity. The resulting surge triggered a domino effect, knocking out critical infrastructure supporting our satellite uplinks.

With the games hanging by a thread, the unsung heroes of the Los Angeles Department of Water and Power sprang into action. Like elite first responders at a five-alarm fire, the engineers worked with surgical precision, isolating the fault, stabilizing the grid, and clearing the way for our satellite team to reboot vital systems and restore transmission services.

What threatened to unravel into a global broadcasting catastrophe instead faded into a fleeting footnote; a momentary lapse in an otherwise flawless production.

THE DYNAMIC WORLD OF SATELLITE BROADCASTING

In satellite broadcasting, moments like these aren't uncommon; they're simply part of the terrain. This is where our odyssey begins: a heart-pounding, high-stakes world of live satellite transmissions, where chaos is ever-present, and failure is never more than a heartbeat away.

These are the untold stories of daring ideas, groundbreaking innovations, and the visionaries who turned ambitious dreams into global realities. Satellite technology didn't just revolutionize communications; it transformed how we connect with the world and witness history unfold in real time.

CHAPTER 2

THE SATELLITE EVOLUTION

FROM DREAMS TO REALITY

In the autumn of 1945, as the world emerged from the darkness of World War II, a young British scientist and science fiction writer named Arthur C. Clarke sat at his desk, scribbling ideas that would one day change the world. To many, he was just another dreamer, weaving tales of space travel and futuristic technology, but to those who truly listened, Clarke was something more: he was a prophet of the space age.

His idea, outlined in a short technical paper titled "Extra-Terrestrial Relays," was deceptively simple yet breathtakingly bold: a network of satellites placed in orbit approximately 22,236 miles above the Earth, where they would move at the same speed as the planet's rotation, appearing motionless in the sky. These "geostationary satellites," as he called them, could be used for instant global communication, replacing the cumbersome reliance on undersea cables and shortwave radio.

In 1945, the world was not ready for Clarke's vision.

When he submitted his paper to *Wireless World,* a respected British journal, many of his peers dismissed it as little more than the creative thoughts of a science fiction writer. Skeptics scoffed at the idea that humanity could launch and maintain objects in space, let alone use them for communication. The very notion of a permanent satellite circling the Earth was viewed as pure fantasy. It seemed like something out of Clarke's own novels rather than the realm of real science.

While Clarke is best known for *2001: A Space Odyssey*, his most groundbreaking contribution was the proposed concept of an artificial satellite positioned in geostationary orbit, which appears to "hover" over a fixed point on Earth. This revolutionary idea laid the foundation for modern satellite technology, enabling everything from live broadcasting to global internet coverage. In recognition of his foresight, the geostationary orbit is often referred to as the "Clarke Orbit," a lasting tribute to the man who turned science fiction into reality.

Now, eighty years later, we embark on a fascinating journey through the evolution of satellite technology, tracing its roots back to Clarke's visionary idea. Arthur C. Clarke didn't just predict the future; he launched it into orbit and dared the world to follow.

SPUTNIK: THE SPARK THAT IGNITED A NEW ERA

Just twelve years after World War II ended, 1957 ushered in a bold new chapter in human history. The launch of *Sputnik 1* marked the birth of the Space Age and ignited a fierce Cold War rivalry between the United States and the Soviet Union. Space was no longer merely a domain of dreams; it had become a battleground for global power.

In an instant, the impossible became reality. The Soviet Union had beaten the United States into space, proving the Earth's final frontier was now within reach. No longer confined to the pages of science fiction, *Sputnik*'s launch heralded a new age, one where the sky was no longer the limit and where communications, warfare, and global influence would soon be dictated by those who mastered the cosmos.

For the United States, *Sputnik* was more than a wake-up call; it was an alarm bell that shook the nation to its core. If the Soviets could launch a satellite into orbit, what else could they launch into space? Missiles? Nuclear warheads? The Cold War, once confined to earthly borders, had just expanded into the vast, uncharted domain of space.

What followed was a surge of innovation, urgency, and fear. The Space Race was no longer just about national pride; it was about survival. The U.S. government scrambled to fund research, scientists worked at a feverish pace, and an entire generation of young Americans was thrust into an era of technological urgency. *Sputnik* became a catalyst that would reshape the course of history.

For many Americans, the launch of *Sputnik* hit like an earthquake, shaking the nation's confidence. The United States had long considered itself technologically superior to the Soviet Union, but *Sputnik* shattered that illusion. Suddenly, there was a growing fear that America was not only falling behind in the space race but, more alarmingly, in the arms race as well.

Fear gripped the nation. Schools conducted "duck and cover" drills, preparing children for an attack that no desk could shield them from. Years later, George Carlin would mock the absurdity of these drills, quipping, "If I see a mushroom cloud on the weather satellite, I'm not going to worry about the rain showers," but in 1957, there was nothing amusing about it. *Sputnik* was a cold slap of reality, jolting the nation out of its postwar complacency and into a new era of vigilance and uncertainty.

Despite the initial shock, *Sputnik* ultimately inspired a determined response. It served as a catalyst that propelled the United States to push the boundaries of space exploration. This relentless drive culminated in the Apollo program and, ultimately, the historic 1969 *Apollo 11* mission, when humans first set foot on the moon.

The launch of *Sputnik* not only altered the trajectory of international relations but also laid the foundation for the modern era of satellite communications and space exploration. It heralded the dawn of the Information Age, changing the world forever.

AMERICA'S BIRTH OF SATELLITE COMMUNICATIONS

In response to the Russian satellite launch, President Dwight D. Eisenhower signed the National Aeronautics and Space Act on October 1, 1958, establishing NASA to lead the nation's effort in space exploration and technological advancement. This historic move galvanized both public and government support for science and education, leading to unprecedented funding for research and development in space exploration and satellite communications.

America's first significant step into satellite communications came with SCORE (Signal Communications by Orbiting Relay Equipment), launched on December 19, 1958. This 150-pound satellite broadcast the first-ever voice message from space, a recorded Christmas greeting from President Eisenhower: "America's wish for peace on Earth and goodwill toward men everywhere."

Telstar 1: America's First Live Transatlantic TV Broadcast

Under President John F. Kennedy, America's space ambitions soared. On May 25, 1961, Kennedy famously challenged the nation to land an astronaut on the Moon before the decade's end. His administration also advanced satellite communications with the Communications Satellite Act of 1962, which led to the creation of COMSAT, a private corporation tasked with developing and managing commercial telecommunications satellites.

The launch of Telstar 1 on July 10, 1962, revolutionized global communications by enabling the first live transatlantic television broadcast. This groundbreaking achievement was a collaboration between AT&T, Bell Labs, NASA, the British General Post Office, and the French National PTT, marking a significant milestone in real-time, worldwide connectivity.

On July 23, Telstar 1 transmitted history's first live intercontinental broadcast, featuring remarks from President Kennedy and breathtaking images of the American flag and the Statue of Liberty during a twenty-minute presentation.

More than a technological revolution, Telstar 1 proved the viability of satellite-based international communications and paved the way for geostationary satellites, which would become the backbone of modern global connectivity.

The Rise of Geostationary Communication Satellites: Syncom 3

On August 19, 1964, NASA launched Syncom 3, a small but ground-breaking marvel, the world's first operational geostationary communications satellite. Suspended 22,236 miles above the vast Pacific Ocean at exactly 180° west longitude, Syncom 3 held a perfect, unwavering position relative to Earth. Like a silent sentinel, it enabled continuous communication across an immense expanse of the globe. This was no longer mere theory; it was the realization of a visionary idea first imagined nearly twenty years earlier by Arthur C. Clarke.

Within weeks, Syncom 3 demonstrated its revolutionary power by carrying the first live television broadcast from Asia to the United States during the 1964 Tokyo Summer Olympics. Across America, millions watched the opening ceremony, and I was among those glued to the flickering black-and-white screen as this moment unfolded, forever changing how we experienced history.

In the 1960s, the Olympics were more than sport; they were a stage where two superpowers battled for global hearts and minds. For Americans like me, each race, dive, and finish wasn't just about medals; it was a symbol of national pride amid Cold War tensions. The Soviet Union and the United States faced off not only on the field but in a silent war of ideology and technology.

Amid the roar of the crowd, NBC announcer Bud Palmer's voice cut through with a revelation that electrified the moment: the Olympic Games were being broadcast live via satellite for the first time. Palmer painted a vivid picture of the signal's incredible journey, from the cheering stadium in Tokyo, soaring up to a silent, gleaming satellite crafted by Hughes Aircraft Company thousands of miles above Earth, then racing back down to flickering screens in millions of American homes.

I was spellbound by that presentation; the impossible was happening before my very eyes. We weren't merely watching the games anymore; we were witnessing human ingenuity bridge continents and time zones, connecting us all in real time. Little did I know then that, just a few years later, I would work for a company associated with that very satellite, becoming part of this incredible era of technology and communication.

Satellite Communications on the World Stage

Also, in 1964, the United Nations sparked a global transformation by forming Intelsat, a cooperative dedicated to connecting the world through reliable satellite communications, even reaching the most remote corners of the planet. Just a year later, Intelsat catapulted this vision into reality with the launch of Early Bird (Intelsat I), the first commercial satellite placed in geostationary orbit.

For the first time, live broadcasts, international phone calls, and data transmissions zipped effortlessly across continents. Spearheaded by Hughes Aircraft and supported by COMSAT and Intelsat, this pioneering satellite cemented the United States as a leader in space technology.

Early Bird laid the groundwork for today's digital age, powering everything from high-speed internet and HDTV to the countless cable channels we now take for granted. Its success opened the door to the modern era of global satellite communications, forever changing the way humanity stays connected.

The success of *Telstar 1*, *Syncom 3*, and *Early Bird* not only pushed the boundaries of technology but also played a critical role in the space race with the Soviet Union. These milestones solidified America's place at the forefront of satellite technology, showcasing its unmatched technological and communications expertise on the global stage.

AMERICA'S QUEST

In the aftermath of *Sputnik*, the United States felt a deep sense of urgency. The space race was as much about proving America's place in the world as it was about technology. Enter NASA, formed to lead this charge and show the world that America could still reach for the stars.

Among the first to answer the call was John Glenn, a courageous pilot ready to do what had never been done before. In 1962, Glenn became the first American to orbit the Earth, giving America a victory and a renewed sense of hope. Glenn was more than an astronaut; he embodied America's resolve to rise, compete, and ultimately lead the world in space exploration.

But the actual test still lay ahead: the Moon. The stakes were higher than ever. The world watched as President John F. Kennedy set the audacious goal to land a man on the Moon before the decade was out. NASA's Apollo program faced countless setbacks and near-disasters, but it pressed on. Then, on July 20, 1969, Neil Armstrong and Buzz Aldrin made history. The *Apollo 11* mission landed them on the Moon, a feat many thought impossible.

It wasn't just a victory for America; it was a triumph for all humanity. The Moon landing proved that with unwavering determination and bold vision, there were no limits to what could be achieved. From the relentless hum of *Sputnik* to Armstrong's iconic step, America had not only reclaimed its place among the stars and reignited the dreams of the entire world. The impossible had become reality, and in that moment, humanity stood united in the belief that anything was possible.

This story of space exploration and satellite technology is one of visionary, groundbreaking achievements and the relentless pursuit of innovation, all aimed at connecting the world. In the 1960s, the TV show *Star Trek* captured this spirit of exploration with its iconic opening line: "Space: the final frontier . . . where no man has gone before."

This catchphrase echoed America's aspiration for space exploration and satellite communications, inspiring a generation to dream bigger and reach for the stars.

APOLLO 11 MOON MISSION

On July 16, 1969, millions of Americans gathered around their televisions, holding their breath as the Saturn V rocket stood tall at Kennedy Space Center. The anticipation was electric. This was it, the moment when a dream that had seemed impossible for centuries was about to become reality. *Apollo 11*, carrying Neil Armstrong, Buzz Aldrin, and

Michael Collins, was destined for the Moon. As the countdown reached zero, a deafening roar filled the air as the rocket ignited, shaking the very ground beneath it and sending the crew toward the stars.

For three days, the nation remained spellbound as the astronauts hurtled through the silent void of space. Then came July 20, a night unlike any other. Streets were eerily quiet, bars and restaurants stood empty, as families huddled around black-and-white televisions, watching a ghostly image beamed back from the Moon's surface. Then, at 4:17 p.m. EDT, Armstrong's steady voice crackled over the speakers: "Houston, Tranquility Base here. The *Eagle* has landed."

A collective sigh of relief swept across the nation, but nothing could prepare America for what happened next. The *Eagle's* hatch opened, and the world watched breathlessly as Armstrong slowly descended the ladder. His boots hit each rung, the suspense building, until he reached the last step. The world seemed to pause as his boots pressed into the alien dust. Then came the words that would resonate throughout history: "That's one small step for man, one giant leap for mankind."

In that instant, people erupted into cheers, overcome by the sheer magnitude of what they just witnessed.

As Aldrin descended to join Armstrong, the two astronauts began their exploration of the Moon with deliberate, measured steps, floating lightly across the unfamiliar terrain. In a moment rich with symbolism, Aldrin unfurled the American flag, and, together with Armstrong, anchored it to the surface near the lunar module. Before departing, they left behind a plaque that read: "We came in peace for all mankind."

Broadcasting from the Moon

Apollo 11 was more than a mission to the Moon; it was a milestone in broadcasting. For the first time, humanity watched live images from another world as Neil Armstrong and Buzz Aldrin stepped onto the lunar surface. Their every move was captured by a black-and-white RCA camera, specially designed for space.

The signal, converted into radio waves by the Lunar Module's communications system, raced through space at 186,282 miles per second via the S-band. After traveling nearly 239,000 miles, it was caught by NASA's Deep Space Network antennas at Goldstone, California, and in Australia at Honeysuckle Creek and Parkes. These massive dishes amplified the faint signal, allowing Armstrong's first words on the Moon to be relayed back to Earth.

From there, the footage was converted into standard broadcast format, sent to Houston's Mission Control, and retransmitted to television networks worldwide via satellite. In total, the signal traveled more than 330,000 miles before reaching living rooms across the globe, arriving with only a two-second delay. Even with an additional satellite hop to Europe and Asia, the images appeared nearly instantly, transforming a singular achievement in space into a shared triumph for all humanity.

REKINDLING THE AMERICAN SPIRIT OF EXPLORATION

The *Apollo 11* moon landing in 1969 was a defining moment of American pride, ambition, and ingenuity. The sight of Neil Armstrong and Buzz Aldrin planting the Stars and Stripes on the lunar surface symbolized the limitless potential of a nation driven by courage and determination. For millions of Americans, the Apollo missions proved that the impossible could be conquered, igniting a sense of unity and national pride that transcended generations.

However, after decades of shifting priorities, America's presence in deep space waned. Manned lunar missions became a distant memory, and the spirit of exploration that once defined the nation seemed to fade. Now, that spirit has been reignited with renewed purpose. With the establishment of the U.S. Space Force and NASA's Artemis program, the United States is preparing to return to the Moon, and this time to stay.

Artemis, named after the twin sister of Apollo in Greek mythology, will not only land the first woman and first person of color on the lunar surface, but it will also lay the foundation for human exploration of Mars and beyond. The mission is about reclaiming the pioneering spirit that propelled America to greatness. It's about inspiring a new generation to dream bigger, reach farther, and once again believe in the extraordinary potential of the Red, White, and Blue.

As we set our sights on the Moon and beyond, America stands at the dawn of a new space age. The same boldness that sent astronauts to the lunar surface decades ago now fuels a new era of discovery, ensuring that the American legacy in space is far from over. The stars have always been within our grasp, and now we're reaching for them once more.

In keeping with this spirit of progress and inclusivity, *Star Trek: The Next Generation* updated its iconic opening phrase to "Space: the final frontier . . . where '*no one*' has gone before."

This revision honored the evolving values of equality and opportunity, aligning with the American spirit of exploration and the belief that the quest for discovery belongs to everyone.

It was a slight change in words, but a giant leap in meaning, reminding us that the journey into the unknown is not reserved for the few but shared by all who dare to dream.

CHAPTER 3

THE JOURNEY

THE AGE OF AQUARIUS

The *Apollo 11* Moon landing in 1969 was a triumph of science and a turning point for global media and human connection. Powered by satellite communications, it marked the first time the world came together to witness history unfold in real time. No longer reliant on next-day headlines, millions across continents experienced the moment simultaneously. Satellites, once distant, silent sentinels in the sky, suddenly became invisible threads stitching humanity together. On that day, technology and spirit converged: the Space Age merged with the Age of Aquarius, and history wasn't merely reported, it was shared, live and unfiltered.

Leading up to the mission, the Vietnam War raged on. Nixon had just taken the oath of office, and the nation teetered between hope and disillusionment. Then came the Manson murders, shattering whatever innocence still remained. Just days later, Woodstock tried to reclaim it with music, peace, and a generation's yearning for something better.

One song seemed to capture the spirit of the era: "Aquarius/Let the Sunshine In" by The 5th Dimension, a psychedelic-soul anthem wrapped in hope and cosmic consciousness. The world stood at a profound crossroads, its cultural axis tilting under the weight of revolution, innovation, and the restless pursuit of new frontiers.

In a year defined by astronauts walking on the Moon and music that ignited the soul of a generation, I found my true calling, a path that would lead me into a technological awakening. After graduating from Franklin Pierce College (now Franklin Pierce University) in April 1969, the Vietnam War raged on like an ever-present storm. Like every able-bodied young man classified 1A, I was a prime target for the draft. That heavy shadow loomed over my future, making employers wary of hiring someone who might be swept away to war at any moment.

One sunlit afternoon, as I played ball, destiny walked onto the field. Between innings, the father of one of my teammates approached and asked, "Still looking for a job?"

Without hesitation, I shot back, "YES!" with every ounce of hope and conviction I had.

What I didn't realize then was that this was no ordinary man. He was the head of the department responsible for all professional sports broadcasts in the United States for Hughes Sports Network (HSN), a subsidiary of Hughes Aircraft Company.

From the moment I started working for him, I realized he was a walking encyclopedia of broadcasting. Before joining Hughes, he had dedicated three decades to ABC Television Network, bringing with him a treasure trove of experience, wisdom, and stories that could fill a lifetime. He became my mentor, with a rare blend of patience, generosity, and passionate commitment to teaching.

One day, driven by curiosity, I asked him, "Why are you so dedicated to teaching me everything you know about broadcasting?"

He smiled, eyes twinkling with a mix of humor and wisdom, and said, "Bob, if I teach you everything I know, it forces me to learn something new so that I can stay one step ahead of you as your boss."

Those words sank deep into my soul. From that moment on, I vowed to make it my creed to share my own knowledge openly with colleagues and associates. His lesson was clear and unforgettable: knowledge isn't something to be guarded—it's a gift meant to be shared.

At the time, HSN was the reigning champion in delivering transmission services for the major professional sports leagues. Before satellite broadcasting took hold, live events were delivered via AT&T's terrestrial microwave and coaxial cable systems: a vast, carefully orchestrated network of relay stations stretching across the country.

Then came a moment that changed everything.

The Apollo 11 Moon landing wasn't just a giant leap for mankind; it was a defining moment in my life. Watching those faint images, carried from the Moon through Hughes-built satellites, my living room became a front-row seat to history. I wasn't just observing; I was witnessing.

For the first time, I understood the true power of television. We experienced the landing together, in real time—millions around the world holding our breath at the same instant, eyes fixed on the same grainy screen. It felt as if we were standing on the lunar surface ourselves.

That night, I realized broadcasting was destined to rise beyond earthly limits. The future would travel on those silent, invisible highways of satellites far above, delivering moments like this to the world. And I knew I wanted to be part of that extraordinary journey.

HUGHES SPORTS NETWORK

Howard Hughes was many things—a Hollywood mogul, record-setting aviator, and eccentric recluse—but behind the legend lay a legacy few recognize. While reporters probed his private life, Hughes was quietly reshaping the future of global communication through Hughes Aircraft Company.

Among Howard Hughes's many audacious ventures, one of his most quietly transformative moves was the acquisition of Sports Network, Inc. To the outside world, it seemed like just another business deal. But for those of us in the industry, it felt like the final systems check before liftoff. Broadcasting was about to launch into a new era.

When Hughes rebranded the company as Hughes Sports Network (HSN), he unveiled a vision few could yet grasp. Sports, he understood, weren't merely games; they were a global force waiting to be unleashed. With the right blend of technology and strategy, Hughes believed sports could break down borders and captivate audiences around the world.

Back then, every broadcast had to navigate the sprawling, labyrinthine maze of the AT&T terrestrial network, a vast web of coaxial cables and microwave towers stretching across the country. From the late 1940s to the 1970s, American Telephone and Telegraph Company (AT&T) maintained a strong grip on broadcast transmissions.

Their microwave relay network, a marvel of engineering in its day, relied on towers spaced roughly thirty miles apart, each one a critical baton in a high-stakes relay race. Because microwave signals required an unbroken line of sight, even the slightest interference—such as a passing storm, a flock of migrating birds, or a minor equipment glitch—could send an entire broadcast spiraling into chaos.

Coordinating domestic broadcasts was a challenge in itself, but arranging an international transmission was an entirely different beast. In the 1970s, securing satellite time for a live global event was the broadcasting equivalent of a moonshot, costly beyond belief, logistically mind-bending, and tangled in red tape and bureaucracy so dense it made the IRS look like a model of efficiency.

And yet, in the icy shadow of Cold War tensions, one broadcast cut through the silence, defying expectations and reshaping history, when Nixon revealed his intention to visit the people of China.

NIXON'S VISIT TO THE PEOPLE'S REPUBLIC OF CHINA

On July 15, 1971, President Richard Nixon's televised announcement sent shockwaves across the globe: the United States had reached an agreement for Nixon to visit the People's Republic of China the following year. It was a bold, unprecedented move that broke with decades of Cold War isolation.

For years, the U.S. had refused to recognize China's communist government. Nixon's decision, however, sent a powerful message: America was ready to turn a new page in international diplomacy. The visit would go down as one of the most significant diplomatic breakthroughs of the twentieth century, not to mention a technical milestone. For the first time in history, a live television audience would witness the opening of a new chapter in global relations as it unfolded, beamed via satellite across continents and time zones.

Coordinating that coverage required a fusion of technical precision, real-time problem-solving, and the bold application of space-age technology to bring the world closer together. It was a mission steeped in urgency and complexity; a defining moment in the evolution of global broadcasting.

NBC was staring down a once-in-a-century challenge: how do you broadcast live from a nation that had been sealed off from the West for decades? Nixon's stunning announcement of his upcoming visit to China was still reverberating through newsrooms when NBC was tasked with the assignment. They would be the "pool feed" broadcaster, responsible not only for their own coverage but for delivering the first televised images to every major U.S. network. The mission was clear: the world had to see Nixon's first steps onto Chinese soil. The question was: how?

The very next morning, my phone rang. On the line was NBC's broadcast department, their voices edged with urgency. They needed answers and fast. China's telecommunications network at the time was little more than a relic: fragile, inconsistent, and nowhere near capable of transmitting a live signal beyond its borders, let alone to an international satellite.

That's when I turned to Hughes Aircraft Company, the masterminds behind satellite technology and the very architects of America's space-age communications. Hughes engineers didn't simply offer suggestions; they reimagined the impossible. Their bold solution: construct a functioning satellite gateway on Chinese soil, link it to the global Intelsat system, and deliver history to the world in real time.

The blueprint had the intrigue of a Cold War spy thriller. An entire earth station, disassembled into transportable panels, would be shipped into China piece by piece, along with five tons of state-of-the-art electronics. Some would travel by air, others by sea; whatever it took to get the gear in. Once on site, the team would reassemble it with surgical precision. Every bolt, every cable, every single test module had to survive the journey intact. There was no plan B. One missing component, one faulty connection, and the entire mission would collapse like a house of cards.

THE BROADCAST THAT CHANGED THE WORLD

With the clock ticking and the world watching, Hughes Aircraft pulled off what many later called a technological miracle. By the time Air Force One touched down in China in February 1972, a new ground station was fully operational, ready to deliver one of the most significant diplomatic moments directly into American living rooms.

President Nixon would later call it "the week that changed the world," and for good reason. The broadcast redefined international television coverage, demonstrating that satellites could bridge even the most profound ideological divides. It was not only one of the earliest uses of global satellite transmission for diplomacy, but also China's first step onto the stage of modern communications.

For millions of Americans, watching Nixon's visit to China unfold live was unprecedented. The Cold War had long cast China as an impenetrable adversary. Yet, suddenly, viewers saw their president face-to-face with Chairman Mao Zedong and Premier Zhou Enlai; a glimpse inside a nation sealed off from the West for over two decades.

The images were unforgettable: Nixon walking atop the Great Wall, touring the Forbidden City, and standing in Tiananmen Square beneath Mao's massive portrait. Each moment carried symbolism that transcended words, signaling a new chapter in global relations.

The breakthrough wasn't only in diplomacy, but also in the skies above. For the first time, a live broadcast crossed the Bamboo Curtain, defying both political and technological barriers. In that moment, the world stood as one, witnessing history carried not by diplomats or armies, but by satellites, transforming barriers of isolation into bridges of communication.

THE MAN WHO LAUNCHED THE SATELLITE COMMUNICATIONS ERA

In 1963, Syncom, a satellite built by Hughes, slipped gracefully into orbit and achieved what no machine had ever done. Suspended high above Earth, it moved in perfect harmony with the planet's rotation, appearing motionless in the sky. For the first time, voices and television images leapt effortlessly across oceans, free from static and delay. The breakthrough was demonstrated the following year, when Syncom carried live broadcasts of the 1964 Tokyo Olympics, allowing audiences around the world to experience the games as they unfolded.

Then came the defining leap. On April 6, 1965, Hughes engineers launched Intelsat I—Early Bird. Though no larger than a household refrigerator, it carried the weight of history, opening the door to the first live commercial transatlantic broadcasts and real-time international phone calls. It was a milestone that permanently reshaped how the world connects.

Hughes never stepped forward to claim the triumph. By then, his eccentricities dominated the headlines. Yet far above the clouds, his vision merged with Arthur C. Clarke's dream of geostationary satellites. Together, imagination and engineering rewrote the way humanity communicates. Hughes's legacy in satellite technology remains one of the most iconic achievements of the twentieth century.

Today, every broadcast, every international call, and every digital stream racing across the globe carries echoes of that breakthrough. Hughes may be remembered as the billionaire recluse, but the satellites he set aloft endure as silent sentinels in the sky, a lasting monument to the man who ushered in the Satellite Communications Era.

Living in Hughes' Shadow

Although I never met Howard Hughes in person, his shadow followed my career. I first felt it as a young viewer of the 1964 Tokyo Olympics, then again during the *Apollo 11* Moon Mission, made possible by Hughes Aircraft satellites. Years later, my path led me to HSN, a

Hughes Aircraft subsidiary, and eventually, when I launched my own company, I became one of Hughes Communications' largest customers.

This connection only deepened when I collaborated with Hughes Communications on the world's first high-definition television transmission for NHK in Tokyo, and later on the first U.S. HDTV broadcast via domestic satellite. But Hughes's influence reached far beyond sharper pictures and technological milestones. I remember visiting Hughes Aircraft's Space and Communications Group in El Segundo, standing in quiet awe before Westar VI and Palapa B2, the first satellites ever retrieved from outer space by NASA shuttle crews.

It's a memory that has stayed with me, a reminder of the man behind the machines and the invisible guiding light his genius cast along the path of my journey.

By the early 1970s, Hughes had all but vanished, retreating into the lavish seclusion of Las Vegas hotel suites, where myth had long eclipsed the man himself. My closest encounter wasn't with a legend—it was with a man whose whims could bend reality across an entire city.

It was 1972. I was arranging transmission services for KLAS-TV, Hughes's Las Vegas flagship station, when the phone rang late one night. The station manager's voice was tight with tension:

"Bob," he said, "you'll need to cancel tonight's scheduled program. He wants to watch *The Outlaw*."

The Outlaw, Hughes's infamous 1943 western, was remembered less for its story than for Jane Russell's scandalous debut in the bra Hughes had personally designed. Now Hughes wanted it back on the air—not for ratings, not for publicity, but simply because he wanted to watch it.

Most of us change the channel when displeased. Hughes didn't need a remote; he changed the broadcast itself.

It wasn't merely an event—it was a revelation. Frail yet unstoppable, Hughes could turn the smallest desire into a citywide command, reminding me that a single mind could wield influence in ways most could scarcely imagine. In that quiet, humbling moment, I recognized the enduring imprint of a man whose genius reached far beyond the machines he created.

CHAPTER 4

THE DOMESTIC SATELLITE REVOLUTION

THE GEOSTATIONARY SUPERHIGHWAY

In 1945, Clarke envisioned a global network of geostationary satellites; two decades later, Intelsat I ("Early Bird") became the first commercial communications satellite, ushering in space-based global communication and the dawn of the Information Age.

But this was only the beginning.

By the early 1970s, a new idea emerged: if satellites could span the globe, why not deploy them to serve specific nations or regions? In North America, this concept gave rise to the North American Geostationary Orbital Arc, a carefully coordinated section of orbital space allocated to satellites serving the United States, Canada, and Mexico.

With every launch, a new lane was added to the celestial highway. Satellites such as *Anik A1, Westar I, Satcom 1,* and *Comstar 1* expanded the North American Geostationary Orbital Arc, stretching from 61 degrees to 157 degrees west longitude, transforming it into a critical corridor for television and data transmission.

Canada led the way with Telesat's *Anik A1,* launched on November 9, 1972, and positioned at 107.3 degrees west longitude, providing nationwide coverage. Meanwhile, the United States adopted the "Open

Skies Policy," welcoming private enterprise and accelerating innovation in space-based communications.

Clarke's original vision of three orbiting satellites has since evolved into a sprawling, high-capacity network. Together with the North American Geostationary Orbital Arc, it has become a geostationary superhighway of information. Though unseen, these satellites form the backbone of modern communication, carrying everything from live television and the internet to emergency alerts and military transmissions. What once seemed like a vision of the future has now become an invisible network that powers our global communications.

THE UNITED STATES' OPEN SKIES POLICY AND *WESTAR I* SATELLITE

The 1972 Open Skies Policy revolutionized the U.S. domestic satellite industry. It broke the nation's reliance on terrestrial networks and international satellites by allowing private companies to launch and operate their own systems. This shift spurred rapid advancements in satellite technology, expanding capacity and services. The newly competitive market attracted industry giants like COMSAT, Western Union, RCA, Hughes Aircraft, AT&T, and GTE, all eager to establish their foothold.

In response, the FCC approved three pioneering domestic satellite communication systems: Western Union's *Westar I*, RCA's *Satcom 1*, and COMSAT's *Comstar 1*.

This transformation began in earnest on April 13, 1974, when *Westar I* launched aboard a Delta 2914 rocket. As it settled into geostationary orbit at 99 degrees west longitude, the age of domestic satellite television had officially begun. *Westar I* became a cornerstone of modern telecommunications, enabling near-instant transmission of television and data across the United States, a quantum leap forward that brought the nation closer together and became the first domestic satellite in the United States.

THE FUTURE OF TELEVISION HAS ARRIVED

In the mid-1970s, a quiet revolution was taking shape high above the Earth's surface. The age of domestic satellite broadcasting was dawning, and with it came the promise of faster, farther-reaching communication. For decades, AT&T had been the gatekeeper of American airwaves, controlling the infrastructure that carried television, telephone, and data across the country, but in 1972, change was on the horizon.

That year, the Federal Communications Commission (FCC) began laying the groundwork for what would soon become known as the Open Skies Policy. By 1974, the message was clear: private companies could now own and operate their own domestic satellites. The skies were, in theory, open to innovation.

However, like many policy shifts, the reality was more complex. Tucked within the newly issued FCC guidelines was a critical restriction: any satellite-based transmission could interface with AT&T's network only once. While this "single-point-of-contact" rule was crafted to preserve competition and prevent monopolization of backhaul infrastructure, it also created a significant logistical dilemma. Most stadiums and TV stations were already hardwired into AT&T's vast network. The question was: if both the origin and destination of a broadcast were AT&T-connected, how could a satellite transmission be routed without violating the rule?

A New Endeavor

After years at Hughes Sports Network, where I had carefully honed my broadcasting skills, in 1975, I chose to leave the safety of familiarity and dive headfirst into an ambitious new venture in Los Angeles, CA.

Taking on the role of senior vice president of sales and operations, I knew I was stepping into uncharted territory. But I was ready to help steer this fledgling enterprise to bold ideas and bigger dreams, toward television's next frontier.

The move to Los Angeles was more than just a career change; it was a calculated risk that weighed heavily on my mind. I knew I'd be stepping into the ring with my former employer, the undisputed heavyweight champion of the sports broadcasting world. The challenge was daunting, but that was only part of the story.

Leaving meant uprooting my wife and our newborn son, hauling our lives 3,000 miles away from the comfort and support of our immediate families. The uncertainty of starting fresh in a sprawling, unfamiliar city loomed large. But most pressing was the knowledge that failure wouldn't just be a professional setback, it could ripple through my family's future, threatening the stability and well-being we cherished.

Yet, despite the risks, something pulled me forward. This was a rare opportunity to take on a senior management role at a rapidly growing company, a chance to spread my wings and expand my impact in the broadcasting industry. It promised more than career advancement; it offered the thrill of being at the cutting edge of innovation in communications, a chance to lead and shape the future.

I couldn't let fear hold me back. The potential rewards far outweighed the risks, and with my family standing by my side, I was ready to embrace the unknown.

And then, my first big break arrived sooner than I could have imagined, thanks to the Texas Rangers Baseball Club. I had just secured the exclusive transmission rights for the Rangers' entire television broadcast schedule, a major victory for any young company. But one game, above all, would be remembered, not for the action on the field, but for the historic event unfolding overhead.

On August 9, 1975, as the Texas Rangers faced off against the Milwaukee Brewers, a quiet revolution was taking place behind the scenes. What appeared to be just another summer ballgame quickly became a historic milestone, the very first live, multi-carrier television transmission via a domestic satellite in U.S. history.

Weathering the Storm of History

The night before the broadcast, I met with Western Union's executives at their Upper Saddle River headquarters in New Jersey. Every detail had been mapped, every contingency considered, every signal traced to perfection. We were ready . . . or at least, I thought we were.

I had scouted every angle and covered every base until Mother Nature stepped onto the mound and hurled a wicked curveball I never saw coming.

The next morning, August 9, Milwaukee lay under siege. Not a gentle drizzle, but relentless, unyielding rain, pouring from a sky so heavy it seemed determined to swallow the city whole. Months of preparation, sleepless nights, and painstaking planning, all of it, threatened to be washed away in an instant.

The skies hung low, a gray veil between us and history. If they didn't clear, our dream of delivering the first live domestic satellite telecast would fade, lost in the fog before the world ever knew it was within reach.

I barely touched my breakfast, my mind already racing ahead to JFK to catch my scheduled flight to Dallas, Texas. I moved through the terminal like a man possessed. By the time the plane landed in Dallas, my nerves were stretched to a breaking point. I made a beeline to KXAS-TV, bracing for whatever news awaited me.

The station lobby was thick with tension. One look at the station manager's face and my stomach tightened. Then, slowly, he broke into a grin.

"Just got word, the skies in Milwaukee are clearing. Game's on."

THE BROADCAST THAT MADE TELEVISION HISTORY

The energy in the KXAS control room was electric. As the broadcast began with the Texas Rangers TV Network's opening announcement, the anticipation reached its peak. Then, broadcaster Dick Risenhoover's voice broke through the airwaves, marking the moment we'd all been waiting for: "This is the first-ever live televised, multi-carrier event broadcast to the American public via domestic satellite."

His words reverberated, a powerful declaration that signaled a groundbreaking moment in television history.

The broadcast signal began its journey at Milwaukee County Stadium, traveling through MRC's private microwave network to Western Union's Lake Geneva Earth Station. From there, it soared 22,236 miles into space via *Westar I*, America's first domestic satellite. The signal then descended to Western Union's Cedar Hill station near Dallas, passed through CPI's private microwave system, connected at AT&T's Television Operating Center, and finally reached KXAS-TV, the Rangers' flagship station in Dallas/Fort Worth.

The journey was nothing short of extraordinary. In an instant, the signal covered more distance than most people would travel in a lifetime, an astounding 45,000 miles in a fraction of a second.

Inside the KXAS control room, two monitors sat side by side, one showing the satellite feed and the other the landline signal. The difference between them was striking. The satellite feed was bright, crisp, and vibrant, like the first rays of sunlight breaking through a morning mist. In contrast, the landline feed, degraded by miles of microwave relay hops, appeared faded and dull.

But the monitors revealed a hidden story. The satellite feed lagged by a quarter of a second, a subtle delay caused by its long trek through space, while the landline signal followed a more direct path. I watched the AT&T feed first: the batter swung, made contact, and sprinted toward first base. Then, I shifted my gaze to the satellite feed, watching the exact same action unfold, but just a beat behind.

Viewers at home would never notice the difference; they saw only one feed. But in the control room, I had the rare privilege of watching both unfold side by side, in real time.

That day, KXAS-TV broadcast the entire game from Milwaukee, using the satellite feed, quietly ushering in a new era in television broadcasting. To the average viewer, it may have seemed like just another Saturday afternoon baseball game. But the vivid clarity, rich colors, and stunning sharpness of the picture told a different story, one of innovation and change.

We knew it wasn't just another routine broadcast. August 9, 1975, wasn't simply another date on the calendar. It was the day television soared into the skies; the day sports broadcasting broke free from the constraints of the past and entered the satellite age.

The Dawn of Satellite Sports Broadcasting

Having just achieved a landmark breakthrough, I could see the next wave of technological advances on the horizon—innovations that would solidify satellite transmission as the future of live sports. They would transform the fan experience forever, drawing audiences into every moment, every cheer, every heartbeat that brought the action to life.

That future surged ahead on October 8, 1975, when, for the first time ever, an NHL game was transmitted via domestic satellite. The Los Angeles Kings faced off against the legendary Montreal Canadiens inside the storied Montreal Forum. KTLA in Los Angeles carried the broadcast, beaming the electrifying action from Quebec all the way to California, marking the first international origination of a live event relayed by domestic satellite to a U.S. broadcast station.

Just two weeks later, on October 23, KTLA shattered another barrier, redefining professional basketball broadcasting. In yet another groundbreaking first, the station aired the season opener between the Los Angeles Lakers and the New York Knicks from Madison Square Garden. It was the first regular-season NBA game ever transmitted via domestic satellite, catapulting basketball into a new era.

With baseball, hockey, and basketball now soaring through the skies to reach fans, a new era of sports broadcasting had arrived. Satellite transmission didn't merely stretch the boundaries of technology; it shattered them, obliterating the old barriers of geography and infrastructure. This wasn't merely a technical upgrade; it was a defining moment, one that I knew would forever change the landscape of sports broadcasting.

TRAILBLAZERS SOARING TO NEW HEIGHTS

The launch of *Westar I* didn't just turn a page in broadcasting history; it rewrote the entire script. With satellites now orbiting high above, communication was no longer earthbound. The future had arrived, and it was airborne.

In 1975, independent television stations, which had long operated in the shadows of the major networks, banded together under the Independent Television News Association (ITNA). Their breakthrough? A bold, daily news service beamed via satellite, delivering footage in real-time to stations across the country. For regional broadcasters, it was a game-changer, finally leveling the playing field.

A year later, in 1976, visionary producer Ray Beindorf took the concept even further. For the nation's 200th birthday, he stitched together a coast-to-coast TV network for *The Great American Bicentennial Celebration*. It was a first-of-its-kind broadcast, uniting stations through satellite links from remote locations and creating a truly national celebration that had never been attempted before.

Founded as a modest UHF station, Christian Broadcasting Network transformed in 1977 when Pat Robertson took it national via satellite. That single decision—distributing CBN directly to cable systems—shattered geographic barriers, proved the power of satellite-fed networks, and helped ignite the modern cable television revolution.

By 1979, the satellite revolution had reached corporate boardrooms. Ciba-Geigy Corporation made history by conducting one of the first global video conferences, connecting participants across four continents in real time. It was a bold leap that redefined the boundaries of corporate communication and hinted at a future where distance was no longer a barrier.

Here are their stories:

1) News By Satellite: The Reese Schonfeld Story

Meet Reese Schonfeld, a no-nonsense newsman with an uncanny ability to see around corners. I first met Reese in the early 1970s in New

York, when he was rising through the ranks at UPI Television News. Even then, he was less interested in how news had been done than in how it could be delivered faster, reach further, and be freed from film and fixed schedules.

At the same time, my world revolved around satellites. I was arranging transmission services for major international events at Hughes Sports Network—Muhammad Ali prize fights, World Cup soccer, and other live global broadcasts where there was no margin for error. Reese knew my background, and in 1974, during one of our conversations, he confided that he was ready to move on. He wasn't just looking for another position; he was envisioning a full-time, satellite-delivered news service that independent television stations could depend on every day.

After stints at UPI Television News and a brief period at Television News Inc. (TVN), Reese recognized a gap: independent stations still needed national and international news but lacked the resources of the Big Three networks. In 1975, I moved to the West Coast, yet Reese and I stayed connected, and his vision began to take shape—what would become the Independent Television News Association (ITNA).

This time, the conversations were different. Reese had the editorial vision. I had the practical experience of making satellite distribution work in the real world. Together, we focused on the logistics, securing domestic satellite capacity, scheduling transmission windows, coordinating uplinks, and ensuring reliable delivery to stations across the country. ITNA would be a cooperative: stations would contribute stories and receive a shared daily news feed, all carried via *Westar I*.

On November 3, 1975, that vision became reality. As managing director of ITNA, Reese launched the first-ever satellite-based news feed for independent television stations, and I was there helping to arrange the satellite transmission services that made it possible. For the first time, local TV stations could contribute and exchange news stories via domestic satellite, forming a national, collaborative newsroom years before such a concept had a name. It was a historic moment, one that predated CNN's twenty-four-hour news model by nearly five years.

A few years later, at the National Association of Broadcasters

Convention in Las Vegas, Reese and I were once again deep in conversation about the future of news. Across the bustling convention floor, we spotted Ted Turner. Without hesitation, Reese raised his voice and called out:

"Hey, Ted—why don't you start a twenty-four-hour news channel?"

Turner paused, gave a slight nod, and kept walking. That casual remark would plant the seed for what became CNN, a network Reese would later co-found with Turner and serve as its first president, forever changing how the world consumed news.

Looking back, ITNA was more than a stepping stone; it was proof that news could be live, collaborative, and nationwide. Reese brought the editorial courage; satellites brought the reach. Together, they transformed the flow of information.

That same instinct to see what was coming before anyone else would define Reese's career. Years later, he repeated the feat with The Food Network, proving that he didn't just anticipate the future of television—he helped create it.

2) The Great American Bicentennial Celebration

As America's bicentennial approached, Ray Beindorf, former CBS executive turned independent producer, conceived an audacious idea as vast and bold as the country itself. He envisioned a television event that would bring Americans together for a grand, unforgettable celebration of the nation's two hundredth birthday.

Thus, *The Great American Bicentennial Celebration* was born, a twelve-hour live entertainment extravaganza that aired on July 3 and 4, 1976.

The event was a logistical marvel, combining both terrestrial and satellite technologies to broadcast from diverse and iconic locations, including Fort Henry, the deck of the aircraft carrier *USS Constellation*, and the Iwo Jima Memorial. While most of the 105-station network relied on traditional landlines, the real magic came via satellite, which seamlessly connected these remote sites into one extraordinary unified national celebration. Against all odds, the event aired flawlessly, reaching more than 90 percent of American households. Ray Beindorf's vision proved groundbreaking, demonstrating the transformative power of satellite technology to unite a nation in real time.

3) Christian Broadcasting Network (CBN)

In late 1974, I was invited to lunch by a man with an idea that sounded bold, even risky for its time. He wanted to launch a full-time cable network, but not for sports or movies or news. This one would be devoted entirely to religious programming.

He laid out his vision with conviction: a national audience, uninterrupted programming, and a belief that emerging cable systems could carry something more than reruns and local access shows. I remember thinking it *was* a great idea—forward-looking, ambitious, and absolutely unconventional. Still, after weighing the risks and the realities of the era, I chose to pass.

The man was Pat Robertson.

On April 29, 1977, Pat Robertson made a move that would quietly reshape American television. He took the Christian Broadcasting Network beyond its roots as a religious broadcaster and launched it nationally—via domestic satellite—allowing cable systems across the country to carry CBN full-time.

The leap from a local UHF station to a nationwide network was nothing short of revolutionary. Cable television was still finding its footing, and few fully grasped what satellite distribution could unlock. Robertson did. He understood that satellites could bypass traditional broadcast limits, delivering programming directly to cable head ends from coast to coast.

That single decision didn't just expand CBN's reach, it helped validate an entirely new model for television distribution. It laid the foundation for what would later become The Family Channel, and ultimately a major player in mainstream cable television.

Looking back, that lunch in 1974 wasn't just a meeting, it was a front-row seat to an idea far ahead of its time. Pat Robertson didn't just build a network; he announced to the world that satellite cable television had come of age.

4) The World Soybean Report

Who would've thought that soybeans could make TV history? Yet, in August 1979, they did exactly that. Sponsored by the Ciba-Geigy Chemical Corporation, *the World Soybean Report* marked a watershed moment in satellite broadcasting. It connected economists and audiences across four continents: Tokyo, Japan (Asia); London, England (Europe); Rio de Janeiro, Brazil (South America); and Atlanta, Georgia (North America) in a live, interactive global simulcast.

Executing this event was no easy feat. With a thirteen-hour time difference and the need for flawless two-way communications, it required weeks of careful planning. While global video conferences are commonplace today, in 1979, this achievement was nothing short of astonishing. The broadcast was a bold demonstration of satellite technology's emerging power to shrink the world and redefine how business and communication could be conducted.

These stories—from ITNA's satellite-fed news service, to the Bicentennial's coast-to-coast celebration, to Christian Broadcasting Network and a soybean summit that linked distant continents—captured the nation's imagination and illuminated the unsung heroes of satellite broadcasting. Their pioneering spirit and fearless embrace of the unknown helped lay the groundwork for the global communications networks we rely on today. They all may not be household names, but their legacy is woven into the very fabric of modern media.

THE MAJOR TV NETWORKS

In 1975, America's three television giants, ABC, CBS, and NBC, stood at a crossroads. For decades, they had relied on AT&T's vast web of microwave towers and coaxial cables to carry live broadcasts from coast to coast. The system was proven, dependable, and woven into the networks' very identity.

Then came *Westar I*, Western Union's new satellite, the first U.S. domestic communications satellite. Suddenly, television signals could leap 22,236 miles into space and back, bypassing AT&T's empire altogether. To the old guard, it felt reckless. Satellites seemed fragile, unreliable, almost like science fiction. A storm, a misaligned dish, or a failed transponder could plunge a broadcast into silence, with no backup and no safety net.

But while the Big Three hesitated, others moved boldly. Independent stations and fledgling cable programmers saw satellites not as a risk, but as liberation. They could reach hundreds of outlets overnight, without paying tribute to AT&T. For them, satellites weren't a gamble; they were the future, a chance to rewrite the rules of television from the ground up.

CHAPTER 5

THE TRANSFORMATION OF TV BROADCASTING

THE RISE OF CABLE PROGRAMMING

In the mid-1970s, a quiet transformation was underway, one that would challenge the longstanding dominance of ABC, CBS, and NBC, ultimately reshaping the television landscape. The first major shift occurred on September 30, 1975, when HBO ushered in a new era by broadcasting the iconic "Thrilla in Manila" boxing match between Muhammad Ali and Joe Frazier via the *Westar I* satellite. This broadcast bypassed the traditional terrestrial system and delivered content directly to cable viewers, signaling a bold departure from conventional broadcasting.

Then, in February of 1976, HBO took another decisive move by transitioning from *Westar I* to RCA's newly launched *Satcom 1*, a satellite equipped with twenty-four powerful transponders that covered the entire continental United States. This shift not only provided HBO with a more cost-effective distribution model, but it also ignited a surge in satellite-delivered programming. It was the spark that fueled the rapid expansion of the cable television industry.

While HBO was leading the charge, an entrepreneur in Atlanta saw a different opportunity. Ted Turner, the owner of the independent UHF station WTCG, was determined to transcend the limitations of local broadcasting. In December of 1976, he boldly uplinked WTCG to the *Satcom 1* satellite, giving birth to television's first "Superstation."

Almost overnight, Turner's once-obscure Atlanta station became a national presence in cable homes across the country. Renamed WTBS, the channel quickly became a household name, offering a mix of classic movies, sitcom reruns, and live broadcasts of Atlanta Braves baseball games. Turner had transformed a local UHF station into a national powerhouse, proving that satellite technology wasn't just an innovation, it was in fact a full-blown broadcasting revolution.

The impact was swift and profound. The television landscape was soon reshaped by a surge of specialized channels, live sports coverage, and twenty-four-hour news networks. What had once been a regionally constrained medium evolved into a dynamic, nationwide force, forever changing how audiences accessed and experienced entertainment. This wasn't merely a technological leap; it was a seismic cultural shift that ushered in the modern era of television and laid the foundation for the vast programming diversity we now take for granted.

The Birth of Cable Networks

HBO's pioneering success flung open the door to a new realm in television, sparking a wave of creative innovation across the cable industry. A new generation of specialized networks emerged, each offering distinctive, targeted content that brought fresh voices and formats into American homes. Whether focused on sports, news, music, or niche entertainment, each genre found its own space to flourish, setting the stage for the rise of several transformative cable channels, such as: CBN (Christian Broadcasting Network), ESPN (Entertainment and Sports Programming Network), CNN (Cable News Network), and MTV (Music Television).

These groundbreaking channels, among others, introduced targeted, niche programming that resonated with specific audiences. In doing so, cable television redefined viewing habits, ushering in an era of specialized content that continues to shape the modern broadcasting landscape.

RCA *SATCOM 1*: DAWN OF A NEW ERA

The day RCA's *Satcom 1* roared into space, it carried the weight of history and the blueprint for dismantling television's old guard. For decades, three networks—ABC, CBS, and NBC—held the keys to America's screens, deciding what stories were told, when, and how. But *Satcom 1* cracked that monopoly wide open. Overnight, a new wave of entrepreneurs, innovators, and risk-takers could bypass the gatekeepers and beam their vision straight into living rooms from coast to coast.

The old rules were crumbling. Satellite television didn't just make delivery faster or clearer; it blew apart the very notion of a single, centralized media voice. Viewers who had passively accepted whatever the networks served up suddenly realized they had choices. And they loved it. Television was no longer a one-way street. It had become a sprawling frontier, a Wild West of programming where anything could happen and anyone could stake their claim.

By the dawn of the 1980s, the shift was undeniable. The public, tired of network predictability, spun the dial to a new uncharted territory: cable. It was raw, untamed, and electric. On one channel, you might find an indie film that would never touch a network schedule, on another, a no-holds-barred live concert or an unfiltered sporting event. Every flip of the channel was an adventure, and that was part of the thrill.

Cable became the playground for ideas too bizarre, too targeted, or too daring for prime-time television. Late-night comedy pushed boundaries. Reality-based programming started to take shape. Twenty-four-hour news rolled on without pause. Music television turned pop stars into global icons overnight. It shattered the mold of the old world.

This was the beginning of a cultural revolution. Television had outgrown its role as a commercial product; it had become a blank canvas. And with satellites lighting the way, a new generation of storytellers and entertainers seized the opportunity to craft entire worlds from imagination alone. The screen was no longer a window; it had become a portal. Satellite technology had flung open the door to an era where anything could be broadcast to anyone, anywhere, at any time.

THE EROSION OF THE BIG THREE

By the late 1970s, the broadcast model was showing its age. Satellites were rising, cable was exploding, and antennas sprouted on station rooftops and cable head ends across the country. With the launches of *Westar I* and RCA *Satcom 1*, broadcasters suddenly had a way to bypass AT&T's costly web of towers and cables. For the first time, the networks no longer controlled the entire pipeline, and viewers began to realize they had real choices.

Independent stations jumped first. Satellites offered reach, savings, and independence from "Ma Bell." But the Big Three—ABC, CBS, and NBC—moved cautiously. For decades, their empires had been tethered to AT&T's long lines. That infrastructure was expensive, temperamental, and far from perfect, yet it was proven. Satellites, by contrast, looked risky, space-age marvels too fragile to bet prime time on.

And yet, the tide turned. Satellites delivered cheaper, cleaner, coast-to-coast signals, and the proof piled up. Slowly, reluctantly, the Big Three followed. They weren't leading the revolution, but they couldn't afford to be left out of it.

Then came the real shock: advertisers. Cable's targeted programming let them reach exactly who they wanted, and ad dollars began to drift away from the network giants. Suddenly, *M*A*S*H* and *The Mary Tyler Moore Show* weren't enough to guarantee dominance.

The networks fought back with bigger budgets, prestige shows, and cable ventures of their own. But the balance of power had shifted. The age of top-down broadcasting was coming to an end, replaced by a more fragmented, viewer-driven world.

For the first time, the future of television wasn't dictated by three letters; it was shaped by millions of hands holding millions of remotes.

THE TRANSITION

As cable subscriptions soared and viewers gained more choices, the once-mighty broadcast networks confronted a stark reality: their era of dominance was ending. The Big Three faced a choice—resist and decline or adapt and survive. They shifted their mindset: cable wasn't the enemy; it was the future.

ABC invested in ESPN, transforming it into a sports powerhouse. CBS built regional sports networks, while NBC launched CNBC and MSNBC, staking its claim in business and political news. These moves weren't just defensive; they were visionary, turning broadcasters into multimedia dynasties.

The networks also seized financial opportunities in cable. They partnered on new channels, ABC with A&E, NBC with USA, CBS with Showtime, and leveraged dual revenue streams from advertising and cable carriage fees. CNN's *twenty-four-hour* news model forced them to innovate further, spawning round-the-clock services like Fox News and MSNBC.

Regulatory changes, including "must-carry" rules and retransmission consent, secured local visibility and added revenue. By the early 1980s, domestic satellite technology had freed television from its terrestrial roots. Content now swept from coast to coast, cable subscriptions soared, and a new kind of space race ignited, one driven not by the legacy broadcast giants, but by networks battling for reach and influence across an ever-expanding, connected world.

TV EXODUS FROM MA BELL

Even the greatest empires are not invincible. As demand for live, high-quality television surged, AT&T's terrestrial network, once the gold standard, began to falter. Regional outages interrupted key broadcasts. Long-distance signals dulled the picture and sound. Its rigid, wired infrastructure strained under the pace of change. For all its dominance, AT&T faced a stark truth: the future was moving faster than its cables could carry.

Breaking away wasn't easy. AT&T wasn't just a service; it was the backbone of American broadcasting. Generations of stations had relied on "Ma Bell" to deliver programming coast to coast. Satellites, still experimental and untested in many eyes, seemed risky. Could anything really match AT&T's reach and reliability?

Curiosity, however, has a way of eroding certainty.

One by one, local stations installed satellite receiving antennas. At first experiments, then essential tools, they delivered undeniable results: crystal-clear signals, lower costs, and unprecedented control over programming. What began with a few daring outliers sparked a movement. Independent stations no longer needed to rent space on AT&T's costly grid; they could chart their own course.

AT&T felt the ground shift beneath it. The empire it had ruled with unquestioned authority was slipping away. By the late 1970s, its stronghold was under siege. By the early 1980s, the exodus was in full swing. Local affiliates, independent broadcasters, and even the Big Three networks embraced satellite technology, not as a novelty, but as the new standard.

AT&T, the longtime gatekeeper of television transmission, found itself on the outside looking in. By the early 1990s, its landline-based TV system had become a relic, a reminder of a bygone era in the relentless evolution of broadcast technology.

CHAPTER 6

SATELLITE RESELLERS AND THE FROST/NIXON INTERVIEWS

SATELLITE RESELLERS: AMBASSADORS OF THE INDUSTRY

By the late 1970s, the television industry was undergoing a seismic transformation. What had once been a vast and open frontier in space was quickly becoming crowded, as TV stations, cable operators, and major networks competed fiercely for limited satellite capacity. The launch of the first domestic communication satellites earlier in the decade had revolutionized television distribution, but in just a few short years, demand had already outstripped supply.

For independent TV stations, satellite technology was a heaven-sent breakthrough. Freed from the crushing cost and logistical headaches of terrestrial relay and microwave systems, they eagerly embraced the new era. Television receive-only (TVRO) antennas, those large, iconic satellite dishes, began sprouting up in parking lots and behind broadcast studios across America, symbolizing a rapidly changing broadcast landscape.

This surge in demand paved the way for a new and essential player in the broadcasting ecosystem: satellite resellers. These innovative companies purchased transponder capacity in bulk from satellite operators and resold it to broadcasters and other clients on a flexible, on-demand basis. The model was a game-changer, providing a cost-effective alternative to full-time leases, ideal for live sports coverage, breaking news, and temporary or ad hoc satellite TV networks.

In a fast-moving, ever-changing marketplace, the companies I worked for emerged as leaders in satellite television transmission services. Their clients were as diverse as the programming they carried, serving local stations, network affiliates, news outlets, corporations, religious broadcasters, and educational institutions alike.

These satellite resellers became key players in this rapidly growing industry, acting as the satellite operators' most effective ambassadors. By offering comprehensive "one-stop shopping," they streamlined access to satellite services, marketing, selling, and facilitating transmissions with remarkable agility.

Beyond sales, resellers played a critical role in live event broadcasting, coordinating transmissions, arranging uplink and downlink facilities, securing production services, and providing expert consulting on video, audio, and data transmissions.

By bundling satellite and terrestrial services under a single umbrella, resellers became indispensable to the industry's growth, fueling the rapid expansion of global communications and enabling a new era of television distribution.

The Pricing Strategy that Transformed the Satellite Industry

In the early days of the satellite era, Western Union clung to a legacy pricing model borrowed from AT&T: a rigid, distance-based fee structure that charged based on both distance and time.

The problem? Satellites don't operate like terrestrial networks. Signals beamed through space aren't bound by geography. Yet Western Union insisted on applying outdated terrestrial logic to a technology designed to transcend distance. Not only was it inefficient, but it also actively discouraged broadcasters from embracing a breakthrough capable of revolutionizing the industry.

Recognizing the flaw for what it was, I championed a bold yet practical solution: a flat-rate pricing model, completely untethered to distance, with satellite time available for purchase in bulk at discounted rates, then resold to broadcasters at a modest markup.

That change proved to be a tipping point, igniting a seismic shift across the broadcasting landscape. Suddenly, broadcasters had access to affordable, flexible satellite capacity, free from the financial strain and bureaucratic red tape of dealing with slow-moving, monolithic carriers. Satellite resellers quickly carved out a profitable niche, positioning themselves as essential intermediaries between satellite operators and an ever-expanding universe of content providers.

For Western Union, long weighed down by an outdated and un-sustainable pricing model, the transformation was profound. By selling significant bulk-buy satellite time commitments to resellers, it revitalized its satellite venture, turning it into a scalable and profitable business.

The result? A win-win-win scenario: broadcasters gained access to new resources, resellers uncovered fresh opportunities, and satellite operators unlocked significant revenue growth.

THE RISE OF SATELLITE PROGRAM DISTRIBUTION

A broadcasting revolution was brewing, one that would upend the long-standing dominance of the three major networks. In their place rose a new wave of syndication, independent producers, and content distributors, all eager to harness the untapped potential of satellite technology. This shift didn't just challenge the status quo; it shattered it.

Satellite transmission proved to be a true game changer: slashing distribution costs, dramatically improving picture quality, and vastly extending the reach of programming. Suddenly, the gates of program distribution were thrown open, unleashing a surge of independent and ad hoc satellite-delivered content that bypassed the traditional network stronghold.

Early adopters saw what the Big Three had missed: an opportunity to break free from antiquated, expensive distribution chains and deliver content directly to audiences.

The key enabler? Satellite receive antenna installed at TV stations, which allowed them to leapfrog costly AT&T long lines and plug straight into the sky.

The results were immediate and powerful: a surge in diverse programming, high-quality transmissions, and an explosion of viewer choice that redefined television. Before domestic satellites entered the scene, national events like the Jerry Lewis MDA Labor Day Telethon relied heavily on AT&T terrestrial landlines and leased private circuits to weave together a national audience. Satellite relay was typically reserved for far-flung destinations like Hawaii.

Everything changed in 1976, when Ray Beindorf's *Great American Bicentennial Celebration* broke new ground. It was the first major special to seamlessly combine domestic satellite and terrestrial links, delivering a coast-to-coast broadcast with unprecedented efficiency. Then, the following spring, television history was made once again, not just for what aired, but for how it reached millions of living rooms across North America with the *Frost/Nixon Interviews*.

THE FROST/NIXON INTERVIEWS

The *Frost/Nixon interviews*, a riveting series of candid and often intense conversations between British broadcaster David Frost and former U.S. President Richard Nixon, became a landmark moment in political television. For the first time since his resignation, Nixon publicly addressed Watergate and his role in it, offering viewers a rare, unfiltered glimpse into the mind of a disgraced leader in search of redemption.

While the interviews held the public spellbound, the behind-the-scenes broadcast effort was just as groundbreaking. None of the major networks wanted to touch it. They dismissed the project as "checkbook journalism,"* a term used to criticize the ethically questionable practice of paying for

* *Checkbook journalism is a term used to describe the practice of paying sources or interviewees for their stories. The major networks had long held to the principle that newsmakers should not be paid. The approach could raise ethical concerns about the integrity and credibility of the information being reported.*

interviews. Frost's production company, Paradine Productions, had report-edly paid Nixon $600,000, plus a share of the profits. Nixon, drowning in legal fees and stripped of presidential income, needed the money.

With no network backing, Frost and his team chose the bold path of building their own broadcast network from scratch. What followed was a technological and logistical triumph: a hybrid, ad hoc syndication network made up of 165 television stations across the United States and Canada. This patchwork coalition included affiliates from all three major networks, as well as numerous independent stations.

Delivering the signal required a level of coordination rarely at-tempted at the time. Eight different carriers were involved, five relying on traditional terrestrial circuits and three on satellite transmission (two domestic and one international). It was one of the first instances where satellite technology was deployed on such a scale to bypass the traditional television establishment, and it worked flawlessly. The success of this unconventional distribution model proved that, with the right tools, determination, and a compelling story, even independent producers could challenge the networks dominance.

However, before I could even see the network lineup to map out the transmissions, a strict non-disclosure agreement (NDA) was placed in front of me. Everyone tied to the broadcast had to sign, no leaks, no slip-ups, absolute secrecy until airtime.

Every move was calculated, every detail critical. And then it happened: the interviews hit the airwaves. The impact was immediate. The broadcast ignited the nation, a lightning rod of attention and debate, proving not only the enduring fascination with Nixon but also the raw, transformative power of satellite-driven syndication. But for millions watching, this was more than just television history; it was a reckoning with a man who had lived at the center of power, scandal, and controversy.

Nixon: Behind the Persona

Richard Nixon's journey through American political life was nothing short of Shakespearean, a tale of towering ambition, bitter rivalries, and haunting defeats. By the time he sat across from David Frost, he was no longer just the thirty-seventh President of the United States. He was a man carrying the weight of triumphs and failures that had unfolded in the glare of cameras and the judgment of history.

His first significant blow came in 1960, when he narrowly lost the presidency to John F. Kennedy. That election underscored a harsh reality of modern politics: television had become the new battlefield. In the nation's first televised debates, Nixon appeared pale, tired, and uncomfortable, while Kennedy radiated confidence and charisma. The margin was razor-thin, but the lesson was undeniable: perception on the screen could shape power itself.

For Nixon, television would remain both ally and adversary, a medium that could propel him forward or betray him in an instant. And nowhere would that tension be more vividly displayed than in his collision with David Frost, when the cameras would once again decide how the world saw Richard Nixon.

Nixon agreed to the Frost interviews because he thought he could control the narrative, avoid hostile questioning, make money, and reintroduce himself to the public on his own terms.

In the end, it didn't exactly go according to Nixon's plan. Like any gripping drama, the premiere broadcast threw in an unexpected twist—a technical glitch so perfectly executed, it felt as if fate had scripted it.

Premiere Night: May 4, 1977

The atmosphere crackled with energy as millions of Americans, from living rooms to bars, gathered around the television sets, waiting for the moment that had been promised for months: Richard Nixon, the fallen president, finally facing the questions he had long avoided. Would he admit to his misdeeds? Would David Frost hold his ground and get Nixon to confess?

The control room buzzed with anxious anticipation. Engineers, producers, and directors worked with laser precision. Every switch was flipped with purpose; every monitor was carefully watched with hawk-like attention.

Then, David Frost appeared on screen, calm and poised. His lips moved as he delivered his carefully rehearsed opening lines. And then, nothing. Silence!

In the control room, confusion turned to panic. "We have no audio!" the technical director barked.

In an instant, engineers scrambled, twisting dials, flipping switches, tracing cables, frantically searching for the source of the problem. Somewhere deep within the maze of wires and circuits, the signal had vanished into the void. The seconds ticked away.

The lead audio engineer's hand moved with urgency, flying over the console. With one swift motion, he cranked up the audio. Instantly, Frost's voice pierced through the silence, cutting through the tension and bringing the broadcast to life.

But the damage was done. Viewers were reminded of none other than Rose Mary Woods, Nixon's fiercely loyal secretary, and the infamous eighteen-and-a-half-minute gap in the White House tapes.

The next morning, the newspapers had a field day. One headline quipped: "Was Rose Mary Woods at the Audio Console?" A nod to the now-infamous Watergate tapes.

In those ninety seconds or so, a minor technical mishap became a part of television folklore; a moment that would live on in the drama surrounding the Frost/Nixon interviews: an iconic clash of personalities, history, and redemption.

The public reaction to the four-part, ninety-minute Frost/Nixon series was nothing short of sensational. Airing across several weeks, the series captivated millions, delivering a rare moment in television history when a former U.S. president was directly confronted about his actions.

Viewers, many of whom had been frustrated by Nixon's evasiveness since Watergate, were riveted by his visible discomfort and eventual admission. When he uttered the words, "I let the American people down," it made newspaper headlines nationwide. It was as close to an apology as the nation had ever received. For some, it was vindication; for others, it was too little, too late.

Initially underestimated, David Frost emerged as an unexpected hero. His persistent, calculated approach forced Nixon into corners no other journalist had managed. The press praised his performance while critics who had doubted his journalistic prowess were stunned by his ability to go for the jugular.

The public was left both satisfied and yearning for more. Ultimately, the Frost/Nixon interviews were watched by more than 50 million people, reaching over 95 percent of American households, making it the most-watched political interview in TV history.

The series redefined broadcast journalism, setting a new benchmark for political interviews. To this day, it remains one of the most talked-about televised events in history, proving that sometimes the greatest courtroom is the court of public opinion.

CHAPTER 7

BREAKING BROADCASTING BOUNDARIES

SHATTERING THE NETWORK MONOPOLY

By the late 1970s, ABC, CBS, and NBC were losing their grip on American television. A technological transformation was underway, powered by satellites and television receive-only (TVRO) earth stations, giving cable programmers, independent producers and broadcasters the tools to challenge the old guard.

Events like Ray Beindorf's *Great American Bicentennial Celebration* and the *Frost/Nixon Interviews* proved a new reality: with satellites, well-produced shows could skip the networks entirely, no need for costly AT&T landlines, and beam directly to national audiences.

Suddenly, network affiliates, once chained to rigid schedules, began installing their own satellite dishes. With a turn of the dial, they could choose between network feeds, syndicated shows, or even produce their own programming.

Live sports, religious programming, and first-run syndicated shows exploded onto the scene. Nationwide reach was no longer a network monopoly. With good content, a satellite uplink, and a little initiative, anyone could build their own TV network.

Syndication

TV syndicators flipped the script by embracing satellite technology: no more shipping bulky film reels or videotapes to local stations. Now, with a satellite dish, stations could pull syndicated programs straight from the sky. Distribution became faster, cheaper, and massively scalable.

This breakthrough gave independent stations a real shot at competing with the networks. Talk shows, music specials, news updates, and live sports could now reach any station with a satellite receiver, no more regional barriers or gatekeeping by the major networks.

Innovation was in full swing. In 1977, the Miss Universe Pageant broke with tradition, leaving network television behind in favor of satellite syndication. Portland's KATU-TV beamed its annual Rose Parade to a regional network, showcasing the growing reach of local programming. By the early 1980s, *Entertainment Tonight* transformed entertainment journalism, delivering its nightly newsmagazine to stations nationwide via satellite.

Syndicators now had complete creative control, keeping every advertising dollar, and targeting niche audiences without network interference. *The PTL Club* (1977) became one of the first religious programs to go national via satellite, while *The Merv Griffin Show* (1979) expanded beyond a handful of affiliates to reach homes coast-to-coast.

Solid Gold (1980) brought top-charting artists straight into living rooms across the nation, fueling the rise of independent stations. On February 25, 1979, *This Is Your Life: Muhammad Ali*, hosted by Billy Crystal, beamed live from The Forum in Inglewood, California, reaching an impressive 130 stations nationwide.

The late 1970s and early 1980s unleashed nothing short of a television upheaval. Independent stations and cable operators, sensing the enormous potential of satellite technology, rushed to install earth station antennas on rooftops and backlots across the country. For the first time, they were no longer tethered to the rigid pipelines of the Big Three. With a simple turn of the dish, stations could pull in live feeds, syndicated programming, or even their own productions from anywhere in the nation.

The monopoly of the Big Three was crumbling, not with a whimper but under a flood of new signals cascading from the sky.

A SURGE IN TV RECEIVE EARTH STATIONS

The installation of TV receive-only antennas at television stations and cable outlets in the late 1970s was a game-changer for the satellite broadcasting industry. Before this shift, broadcasters relied on costly and geographically limited microwave networks and physical tape distribution, especially for programming to and from remote locations like Hawaii. By equipping their facilities with earth station antennas, local TV stations and cable operators gained direct access to satellite feeds, unlocking new programming opportunities. This development dramatically increased efficiency, reduced costs, and expanded access to live content, making satellite transmission a far more attractive option.

One of the most significant outcomes of this transition was the rapid expansion of cable television, powered by TVRO earth stations installed at cable outlets. Networks such as HBO, ESPN, and CNN capitalized on satellite technology to bypass traditional broadcast methods, delivering content directly to cable head ends. This new distribution model enabled nationwide reach, fueling the rise of specialty channels and giving audiences more choices than ever before. At the same time, independent and affiliated TV stations, once at the mercy of major networks, gained the ability to air syndicated programming, live sports, and news feeds without network restrictions, fostering competition and reshaping the television landscape.

FCC Regulations

The Federal Communications Commission (FCC) played a crucial role in accelerating the adoption of satellite technology. Regulatory changes made it easier for local TV affiliates to install receive-only earth stations, streamlining the delivery of national programming. These advancements significantly reduced reliance on expensive terrestrial methods and allowed broadcasters to access a much broader range of content.

As a result, satellite transmission quickly became the preferred method for delivering TV programming across the country. Earth stations, once a complex and costly investment, became commonplace in the industry, enabling TV affiliates of all sizes to deliver high-quality content quickly and efficiently, and ultimately revolutionizing the television industry for good.

RISE OF THE BACKYARD DISH: HOME SATELLITE ENTHUSIASTS

In the late 1970s, a grassroots revolution emerged, spearheaded by a new breed of television enthusiasts known as "backyard dish owners." These pioneers, armed with massive C-band satellite dishes measuring eight to twelve feet in diameter, tapped into cutting-edge technology that unlocked a vast, uncharted realm of domestic satellite television. These iconic dishes soon became fixtures in rural and suburban backyards, symbolizing a new era of broadcasting.

At the time, satellite signals were transmitted unencrypted, granting anyone with the right equipment access to an impressive array of programming, from live sports events to premium cable channels, allowing enthusiasts to bypass traditional cable networks. This unprecedented access was a breakthrough for families in underserved regions, bridging the entertainment gap for countless consumers.

However, the technology came with a hefty price tag. This equipment, including the massive satellite dish, receiver, and other components, often costs thousands of dollars, making it a luxury accessible only to tech-savvy enthusiasts and those with deep pockets. Yet, despite the financial barrier, the allure of unrestricted content and the thrill of tapping into satellite feeds captivated a growing community of passionate users.

By the early 1980s, the popularity of backyard satellite dishes had skyrocketed, prompting a response from the television industry. As networks and cable channels realized the financial implications of free access, they began encrypting their signals.

The shift marked a turning point, requiring dish owners to purchase decoders and subscribe to programming packages. This transition curtailed free viewing and laid the foundation for the modern satellite TV industry, transforming a grassroots movement into a structured commercial enterprise.

HAWAII'S ADOPTION OF DOMESTIC SATELLITE SERVICES

Before the advent of satellite technology, Hawaii's remote location in the Pacific Ocean presented a formidable challenge for broadcasters. Television programming from the U.S. mainland was delivered either via costly undersea cables or through physical tapes flown in, creating significant delays and limiting access to timely content.

In the late 1960s and early 1970s, Hawaii began leveraging international satellite services for live transmissions. This technological leap allowed real-time broadcasts of network programming and special events, such as presidential speeches and the Apollo moon landing, but the cost was prohibitively high.

The mid-1970s marked a transformation in Hawaii broadcasting with the advent of domestic satellite technology. In 1978, a local milestone was achieved when TV stations in Hawaii installed a domestic satellite receive earth station, improving access to live programming from the mainland and greatly reducing the cost of satellite time, compared to international satellite rates.

The impact was immediate and profound. Local stations seized the opportunity to expand their offerings. The television station KHON introduced *The Today Show* to Hawaiian viewers, setting a new standard for morning programming. Not to be outdone, KITV launched *Good Morning America* just months later, and KGMB brought live sports and up-to-the-minute news to audiences across the islands.

THE LAND OF THE MIDNIGHT SUN

After helping bring live programming to Hawaii via domestic satellite, my next adventure took me to Alaska. Unlike Hawaii, where TV stations lacked earth stations, Alaska was ahead of the curve, already using satellite technology for government and business. My mission this time was not to introduce satellites, but to convince leaders to launch a second channel, one dedicated to delivering education and healthcare to Alaska's remote "bush" communities.

Alaska, the "Last Frontier," was even more breathtaking than the postcards suggested. Towering snow-capped peaks, endless tundra, and a rugged landscape where locals thrived on hiking, skiing, dog sledding, and kayaking. Me? I hadn't so much as ridden a dog sled, let alone mushed through a blizzard. Yet there I was, in February of 1978, braving the icy cold winds and subzero temperatures to push satellite technology into the frigid unknown.

Cold Winds and Warm Welcomes

My journey began in Los Angeles, and after a brief stopover in Seattle, I boarded a small, four-seat "milk run" plane, a quirky Alaskan staple known for its unpredictable stops in remote towns—my destination: Juneau, Alaska's capital.

By the time we reached Juneau, the sky had already darkened, even though it was barely late afternoon. Instead of retreating to the warmth of my hotel room, I was drawn to the town's lively heart, the Red Dog Saloon.

Inside, rugged men in flannel shirts and heavy boots filled the room, looking as though they had just stepped off a logging crew. Sawdust covered the floor, and the air smelled of pine and untold stories.

Beer flowed like the Yukon River, and laughter crackled like a campfire on a cold night. At one point, a brass bell rang out, and the entire bar erupted in cheers.

"What is the celebration?" I asked the bartender.

"Someone bought a round for the house," he said with a grin. "It's tradition."

After enjoying my drink, I stepped back into the biting wind and made my way to the hotel. Tomorrow, I'd meet with state lawmakers, not to trade fishing tales, but to pitch a bold vision: linking Alaska's most remote communities by satellite.

The technology? A groundbreaking digital method called STRAP (simultaneous transmission and recovery of alternating pictures). It allowed two distinct TV program feeds to share a single satellite transponder, effectively doubling efficiency. This would open the door to educational content, medical resources, and community programming for even Alaska's most isolated corners. It was bold, innovative, ahead of its time, and exactly the kind of solution that Alaska's rural communities had been longing for.

Innovations vs. Cultural Traditional Values

On the morning of a much-anticipated meeting with Alaska's state legislators, I made my final preparations, going over the STRAP equipment with the engineer and confirming the satellite transponder assignment with Western Union's control center. Everything was set. The technology was ready to showcase its potential.

The town hall was filled with curious faces. Local leaders, state legislators, and a handful of residents had gathered in Juneau to witness an extraordinary glimpse into the future.

Inside, I stood before a group whose everyday lives were not shaped by fiber optics or satellite beams, but by the unyielding rhythms of land and sea. I was there to bridge that gap. Armed with a crisp pitch, a solid sense of purpose, and the promise of satellite technology.

The presentation seemed to be a great success at first. The STRAP system worked flawlessly: two distinct video feeds streamed from a single transponder, proving that even Alaska's most distant corners could stay connected to the outside world.

Nods of approval rippled through the room, but at the back, a solitary figure remained unmoving. He was impossible to miss: a state legislator from Nome, where the Iditarod Trail Sled Dog Race ends, just fifty-five miles south of the Arctic Circle. Gone were the blazer and tie typical of a politician, replaced with a fur-lined parka and well-worn mukluks. He radiated the quiet authority of someone who had spent a lifetime reading the land as if it were scripture. Even in silence, his presence commanded attention.

After the demonstration, I approached him. We exchanged pleasantries and small talk. Then, eager to win him over, I launched into my carefully crafted pitch, explaining how satellite technology could bring education, real-time medical assistance, and entertainment to his community during those long, dark winters.

He listened quietly and patiently.

Then he leaned in, a spark of mischief at the corners of his eyes and centuries of wisdom behind them, and delivered a reply that was unforgettable.

"UGG . . . No Mary Tyler Moore. Fish Hunt! Fish Hunt!"

For his people, survival wasn't measured in prime time; it was measured in salmon runs, lessons in hunting, and the ability to endure.

Back at the office, the rejection came quietly. But I didn't feel disappointment.

That day in Alaska wasn't a failure, but an awakening. It redefined my understanding of communication. It's not just about sending signals or transmitting data through satellites. It's about something far deeper: a shared exchange of values, perspectives, and human connection.

Communication isn't just about speaking. It's about listening, really listening.

Even now, all these years later, I find myself returning to that moment: two Americans standing face to face, shaped by entirely different worlds. One raised in the solitude of the tundra, guided by sled dogs and snow-covered horizons; the other raised in the chaos of subways and the shadows of skyscrapers.

Two lives. One country. The same flag. Yet as far apart as Earth and the Moon. And in that instant, our worlds collided.

We did not meet as opponents; we met as equals. Grounded not in words, but in mutual respect.

It was in that quiet space between words that I discovered a lasting truth: real communication is not measured in bandwidth or signal strength. It lives in the stillness, in those rare moments when nothing is said and yet everything is understood. That is where true communication begins. And no technology, no matter how advanced, can ever replicate that kind of power.

CHAPTER 8

THE VISION

REWRITING THE RULES OF BROADCAST TRANSMISSIONS

Every revolution begins with a spark, but in the early days of satellite broadcasting, that spark was barely a flicker in the dark. The industry clung desperately to the safety of landlines, shackled by the slow, cumbersome weight of outdated technology, blind to the world of possibility looming just beyond the horizon. Yet in the minds of a restless few, the future was already ablaze. They envisioned signals soaring across the sky, free from wires and limits. The true challenge wasn't the technology; it was convincing a skeptical world to believe in a vision few dared to imagine.

That spark roared to life in 1981.

Driven by something stronger than fear or comfort, I walked away from the security of a stable job and from the nation's leading satellite reseller to lead a bold new venture: a San Francisco startup that wasn't just chasing the future, it was hell-bent on reinventing it.

I stepped into the role of president at a new subsidiary in Burbank, California. Our mission? To dismantle the old model and rebuild it from the ground up.

Freed from the stranglehold of outdated landlines, we set out to launch a new era, one defined by speed, agility, and resilience. The vision was daring: real-time, on-demand satellite transmission, deployable anywhere and at any moment, completely untethered from the legacy infrastructure of terrestrial facilities.

The skeptics didn't hesitate. "It's impossible," they sneered. "You can't ditch landlines and still expect reliability."

They couldn't see what we saw: a chance to upend the status quo, to rip out the old rules and rewrite them with a pen dipped in pure ambition.

Our clients weren't the usual suspects. We reached out to the overlooked: the religious broadcasters preaching to niche audiences, Fortune 500 giants demanding flawless communication, concert promoters chasing the perfect live feed, and sports organizers determined to bring games to fans everywhere. Each presented unique challenges, pushing us to break new ground.

We delivered tailored experiences. Broadcasting pharmaceutical training for Pfizer, coordinating nationwide corporate meetings for Ford, and streamlining live technical workshops for Hewlett-Packard.

Long before "video conferencing" became a buzzword, we were trailblazers, pioneering the seamless integration of satellite downlinks, audio-visual projection, and multi-site connectivity into a single, unified service.

With every feed we launched, every so-called impossible mission we pulled off, we weren't just transmitting signals. We were making a statement:

The future wasn't on its way. It had already arrived.

THE TRANSPORTABLE NETWORK

In 1982, I wrote an article titled "The Transportable Network"—a piece that explored a daring new frontier in corporate communication: closed-circuit video conferencing. At the time, this wasn't just a novel idea; it was a revolution in the making, offering companies a powerful alternative to the high cost and time-consuming hours of business travel.

The early '80s were an era of mounting expenses. Airfare, hotels, meals, and rental cars were eating into corporate budgets at an alarming rate. Organizations across various industries, including sales, marketing, education, and media, were seeking a more innovative way to connect their far-flung teams. Video teleconferencing emerged as that answer.

By 1982, we had produced over fifty live satellite video conferences, establishing a new benchmark for business communication. With transportable TVRO antennas capturing signals and bypassing the limits of terrestrial facilities, we proved that 'impossible' was just another challenge waiting to be solved.

One standout moment came with Tandem Computers in Cupertino, California. With a nationwide marketing force scattered across the U.S., Tandem needed to roll out a new product line quickly and effectively without draining its budget on airfare and hotel bills. Instead of flying everyone in, they embraced a live satellite video conference.

We rolled a transportable uplink unit onto their corporate complex and broadcast the product launch straight from headquarters to regional offices across the country. Portable TVRO downlink antennas at each site turned offices into live viewing hubs, eliminating the need for expensive terrestrial infrastructure.

The results were dramatic. Tandem saved tens of thousands in travel expenses, avoided the logistical chaos of nationwide scheduling, and delivered a seamless, high-quality broadcast that redefined what corporate communication could be.

But the real breakthrough reached far beyond a single company. Tandem's event demonstrated something greater: that satellite technology had the power to fundamentally transform the way businesses connect, collaborate, and compete. It was more than an experiment; it was a turning point. The transportable network was no longer just an idea; it had become a reality.

From Innovation to Everyday Integration

Back in 1982, video teleconferencing was a high-wire act, satellites overhead, transportable uplinks on the ground, and a full crew of technicians working to keep every signal alive. Today, the same concept plays out with nothing more than a tap on a screen. Platforms like Zoom, Microsoft Teams, and Google Meet have transformed what was once a corporate luxury into an everyday reality, powering conversations everywhere, from Fortune 500 boardrooms and university lecture halls to kitchen-table startups and grandparents whispering bedtime stories over FaceTime.

What began as a clever workaround to reduce the cost and hassle of business travel has evolved into a cornerstone of modern life. It's a vivid reminder of how far we've come, from an era of logistical hurdles and high-tech coordination to a digital world where connection is instant, seamless, and always within reach.

SPREADING THE WORD OF GOD

In the 1980s, a revolutionary force reshaped how the gospel reached the world, not from the pulpit, but from the heavens. Satellite technology, once the domain of corporate meetings and entertainment, became a celestial highway for spreading the Word of God. What began as a practical tool evolved into a spiritual megaphone, projecting faith across continents and into the hearts of millions. This was the dawn of modern televangelism.

At the forefront stood Billy Graham. His crusades, once confined to stadiums, became global gatherings. Through satellite transmission, his sermons were broadcast live to scores of countries, turning each message into a worldwide spiritual event. Millions who once read his words in newspapers could now see and hear him in real time, sharing a collective experience of faith.

Graham wasn't alone. Oral Roberts staged massive satellite crusades from Tulsa's Mabee Center, sending healing and hope to over 200 cities worldwide. Transportable satellite networks, uplinks paired with TVRO downlinks, freed ministries from costly terrestrial infrastructure, giving them a global pulpit in the sky. The signals were crystal clear, the reach limitless, and the message boundless.

Kenneth Copeland Ministries embraced high-tech productions,

while Marilyn Hickey Ministries pioneered live virtual prayer circles, connecting viewers as if they were in the studio itself.

Satellites turned the sky into a sanctuary. From Graham's stirring crusades to Roberts's televised healing services, faith soared on the wings of innovation, opening a new era in which technology and spirituality met, and the message of hope knew no boundaries.

BROADCASTING UNDER PRESSURE

In the high-stakes world of transportable TV receive only (TVRO) satellite services, there are no second chances, no safety nets. You either nail it, or you go down in flames before millions of viewers.

Every broadcast site was a battlefield of logistics and timing. Terrain had to be scouted with military precision. Equipment wasn't simply delivered; it was dragged over mud, gravel, or ice. TVRO earth stations rose from crates and toolboxes, bolted together, calibrated, and aimed skyward with the accuracy of a marksman. Power was hunted down, miles of cable unspooled, and frequencies tested and retested. No two jobs were alike; each came with its own ambush of obstacles, red tape, brutal weather, rogue interference, or hidden flaws waiting to sabotage everything.

Planning was exhaustive, backups essential, but unpredictability was always the wildcard. And when things went sideways, survival depended not on the gear but on the people. TVRO engineers were troubleshooters, fixers, miracle workers, staying cool under fire, bending rules, and improvising with duct tape, ingenuity, and sheer willpower.

Because in live television, every cable is a fuse, every second a ticking bomb, and every glitch a direct hit to your credibility. The brutal truth? The unexpected isn't an accident; it's an executioner, waiting for the smallest mistake. And when chaos strikes, you don't flinch. You plant your feet, stare it down, and fight like hell to keep the feed alive.

Still think live television is all champagne and caviar? Here's the real story of a TVRO transmission network, where grit, guts, and relentless determination are the only things standing between a flawless broadcast and total disaster.

The Divine Intervention

It was designed to be flawless, a high-profile, nationally televised televangelist broadcast destined to reach millions. But pulling it off meant orchestrating a 123-site transportable satellite downlink network, a task far beyond the routine 30-site TVRO setups. It resembled a broadcast in name only; in truth, it was more like juggling spinning plates in a wind tunnel, knowing that a single slip could bring the whole performance crashing down.

Every cable, every frequency, every timing window had to be plotted with surgical precision. Backup systems were layered like armor, redundancies stacked on redundancies, and contingency plans drafted with the detail of a general preparing for battle.

When the broadcast day finally came, the air buzzed with anticipation. Across the country, technicians called in, their voices cutting through the static of radios and phone lines. One by one, massive dishes swung toward the heavens, locking onto satellites with mechanical certainty. Signals flared across our monitors, first a trickle, then a surge. On the control board, green lights blinked alive, spreading across the map until the screen glowed like a constellation stitched coast to coast.

And for a brief, suspended moment, it was perfect. The network pulsed like a living organism, a symphony conducted across thousands of miles, every note in tune.

Then Memphis happened.

Fifteen minutes past the first scheduled check-in, just two hours before going live, silence. No phone call. No technician. No signal. At first, we shrugged it off. With 123 sites, there were always stragglers. But the silence thickened, pressing in like a storm front. Memphis was late.

Protocol kicked in. Calls went out. Radios crackled with nothing but static. The room, once humming with controlled urgency, turned ice-cold with tension. Then, finally, the line clicked alive. A trembling voice broke through:

"High winds . . . the antenna, it's gone. Torn clean off the trailer. It's in a ditch. Total wreck."

He could barely get the words out, choking on panic. And in that instant, the weight crashed down. This wasn't a glitch. This was a disaster.

We had ninety minutes, just ninety minutes, to bring Memphis back. A key hub, one of our largest and most critical audiences, had gone dark.

Adrenaline surged, hearts pounded, and the room snapped into motion. The battle to save Memphis had begun.

First, we located a technician twenty-five miles from the site. After some fast-talking and a dash of desperate persuasion, he reluctantly agreed to head to the venue. For a fleeting, heart-stopping moment, the room exhaled. We had a shot.

Then, barely twenty minutes later, the phone rang again.

It was him. Fear dripped from every word. "The winds . . . they're too strong," he stammered. "It's too dangerous. I can't go any further."

Just like that, Plan B collapsed under the full weight of nature's fury.

Panic threatened to take hold, but quitting wasn't an option. Not in our DNA. We worked the phones relentlessly, chasing every lead, pleading, convincing, bargaining with fate itself.

And then, after a blur of calls, we struck gold: a third operator, farther out, willing to face the brutal winds. He was in.

Still not satisfied, we reached out to a fourth operator. He had an antenna, no electronics, nothing to make it fully functional, but he was willing to try. By now, we were throwing everything at the problem: no playbook, no guarantees, just raw instinct and grit.

Then, forty minutes before airtime, the original operator, the one whose antenna had been ripped away, called back. He was racing to the venue with whatever scraps of equipment he could salvage—a Hail Mary.

And then . . . the moment we'd been holding our breath for: the third operator's voice crackled over the radio. "Five minutes out . . . I can see the venue," he gasped.

Every head in the ops room snapped to the clock. The room froze. A beat of silence stretched across the monitors. After hours of chaos, maybe, just maybe, we were back in the game.

But fate wasn't finished with us.

As the technician eased off the highway, still gripping the phone, it happened.

Screeching tires. A sickening crunch. Then—CRASH.

Silence. The line went dead.

Seconds later, the phone lit up again. His voice trembled, barely audible over the chaos: "I've been hit . . . sideswiped at the exit ramp. The antenna, it's gone. Ripped clean off the trailer. Shattered."

Memphis was dark. Out. Gone.

The ops room held its breath. Eyes glued to screens, the team stared in disbelief. In a single heartbeat, a major market had vanished from the broadcast, claimed by crushed steel and sheer misfortune.

Then came the dreaded call: we had to tell the client. Memphis, the centerpiece of the program, was dark. It was the last thing anyone wanted to report. The room seemed frozen, every second stretching like a lifetime, monitors that should have been alive reflecting only silence.

And then, as if by divine intervention, Memphis flickered to life, an unexpected spark from the heavens above. Relief swept through the room, a fleeting rush of hope that made the impossible feel attainable.

The truth soon emerged. The technician whose antenna had been ripped into a ditch hadn't given up. Shaken but alive, he clawed through the wreckage, salvaging bent brackets, a cracked feed horn, a half-busted toolbox, and his own stubborn will. He drove straight to the venue, dragging the scraps like a soldier hauling ammo to the front lines.

Almost simultaneously, a backup operator arrived. An antenna in hand, no electronics, no introductions, no chatter. Just a look. An unspoken pact.

Together, they dove in. Seconds bled away. The clock was merciless.

Then, like a sign from above, the satellite signal snapped in. Memphis roared to life.

Not just a fix. A resurrection.

Two strangers, against every odd, had pulled off the impossible.

Skill? Luck? Or something far greater? No one could say. All we knew was that, in that instant, it felt like a miracle.

Memphis was a battle against wreckage, time, and the impossible. Two strangers defied fate itself, hauling shattered equipment to triumph and resurrecting the signal. This is the life of the TVRO engineers and technicians who thrive on the impossible, improvising, problem-solving, and performing last-minute miracles. Against every obstacle and the relentless march of the clock, these operators turned near-disasters into flawless transmissions, moments that might have ended in failure but instead became broadcast legend. And none embodied that spirit more than our next story: the tale of the TVRO Ninja Warrior.

THE TVRO NINJA WARRIOR

In Philadelphia, the city of brotherly love, our TVRO operator thought he'd struck gold. His trailer was parked in the perfect spot, lined up with an unobstructed view for the day's live broadcast.

Then came the problem: the trailer sat squarely in the middle of a restricted zone.

Enter the officer. Grim-faced, arms folded tight, radiating the weary authority of a man who hadn't smiled since the Carter administration. His voice cracked like a drill sergeant, and his citation pad snapped open with the finality of a judge's gavel.

The operator pleaded his case. "Sir, there's a live broadcast happening. Thousands of people inside that arena are depending on this signal." His tone was steady, but underneath ran the urgency of a man who knew the clock was ticking.

Officer No-Nonsense didn't blink. He was stone, unyielding, immovable, deaf to reason. The air thickened, time dragged, and then the verdict fell. Like a hammer on steel, the order was given: a tow truck was on its way.

Moments later, the tow truck arrived, and the situation escalated fast. The driver hooked up the trailer, satellite dish still mounted, and began hoisting it into the air. For most people, this would've been the end of the line . . . game over. But not our technician.

Our guy didn't see a problem; he saw a challenge. Channeling his inner action hero, he leapt onto the trailer like a ninja warrior, climbing up as it dangled mid-air. Cool as a cucumber, he grabbed his tools and started adjusting the dish's alignment, eyes laser-focused on maintaining the satellite signal.

Here's our champion, a full-grown man balancing on a swinging trailer, wielding a wrench in one hand, and a protector in the other, fine-tuning an antenna like it was just another typical assignment. The trailer hung suspended like some sort of gravity-defying piñata and yet managed to keep the broadcast signal rock-solid.

Meanwhile, the tow truck driver and the officer stood there, with their mouths wide open. They'd expected outrage, maybe some yelling or cursing, not a live acrobatic performance. The officer finally snapped out of his stupor and growled, "What on Earth are you doing?"

With the calm confidence of someone who's been through worse, the technician replied, "My job, sir."

The officer and tow truck driver were amazed by the technician's determination and dedication. After a brief discussion and a bit of financial persuasion (let's call it a "goodwill donation"), they struck a compromise.

The policeman and tow driver allowed the trailer to remain suspended in mid-air for the duration of the telecast. In a twist of fate, the new elevated position actually improved the signal quality. What began as a disaster turned into an unexpected triumph, proving once again that sometimes, ingenuity and a little cash can turn the craziest, most unpredictable situations into an unforgettable victory.

CHAPTER 9

ROCK 'N' ROLL

THE KING OF ROCK 'N' ROLL: ALOHA FROM HAWAII

Elvis Presley, the King of Rock 'n' Roll, made history in 1973 with the first-ever global satellite broadcast of a concert. Titled *Aloha from Hawaii*, the event was a pivotal moment in Elvis's career as well as a monumental event in television history.

Aloha from Hawaii was broadcast live via satellite on January 13, 1973, from the Honolulu International Center in Hawaii. The concert was broadcast to forty countries and was watched by an estimated 1.5 billion people worldwide, making it the most-watched broadcast of its time. *Aloha from Hawaii is* remembered not only for its iconic performances but also for its role in transforming the live music industry. It was a pioneering moment in satellite broadcasting, setting the stage for future global concerts and live television events.

It also underscored the importance of satellite technology in shaping the future of television and entertainment, allowing audiences to experience live events from anywhere in the world.

THE NIGHT THE ROLLING STONES ROCKED THE WORLD

It was December 18, 1981, at the Hampton Coliseum in Virginia, and the Rolling Stones were electrifying America on their *Tattoo You* tour. But this night wasn't just about the thousands in the arena, it was about millions. For the first time ever, a live pay-per-view rock concert was broadcast via satellite, bringing the band's raw, high-octane performance into homes, bars, and theaters across North America. The event was boldly billed as *The World's Greatest Rock 'n' Roll Party*.

Live satellite transmissions had already transformed sports, but this was different: a full-scale, multi-camera rock concert, raw and unfiltered, streamed in real time. Fans no longer had to wait for radio reports or highlight reels; they were part of the energy, the chaos, the thrill, as it happened. This was the birth of a new era, where music and technology collided to create global moments.

Inside the arena, the atmosphere was electric. Every seat filled, every eye fixed on the stage. When the Stones appeared, timeless and larger than life, the crowd erupted. From the opening riffs to the pounding drums, Mick Jagger prowled the stage like a wildcat, Keith Richards shredded on guitar, and Charlie Watts and Ronnie Wood drove the rhythm with precision and fury. The building practically vibrated with the frenzy.

And then chaos struck. A fan broke through security, charging toward Jagger. For a heartbeat, it was unclear what would happen. Keith Richards reacted instinctively, swinging his guitar like a weapon, taking down the intruder with a single blow, and then seamlessly continued playing, as if nothing had happened. A new legend had just been written.

The broadcast captured it all, unfiltered, raw, and real, bringing the Stones' peak performance into living rooms across the continent. That night didn't just showcase a band; it created a blueprint for a new kind of global entertainment. Live Aid, the Freddie Mercury Tribute, U2's Zoo TV Tour, all traced their roots back to Hampton.

And as if the night weren't already monumental, it was Mick Jagger's thirty-eighth birthday, a fitting backdrop to a night that would rock the world forever.

THE US FESTIVAL: WORLDWIDE FUSION OF MUSIC AND TECHNOLOGY

In the blazing heat of the California desert in 1982, something extraordinary was stirring near San Bernardino. Steve Wozniak, the bearded genius behind Apple, had already ignited the personal computing revolution. Now, he dreamed even bigger: a cultural revolution powered by music, unity, and technology. This wasn't just a concert; it was Woodstock reimagined by a digital pioneer. Pouring millions of his own dollars into the project, Wozniak transformed Glen Helen Regional Park into a sprawling musical utopia: the US Festival.

Four hundred thousand people would converge for a single idea: music and technology colliding to change the world. Wozniak didn't just want entertainment; he wanted connection. Even amid the Cold War, he dared to invite the Soviet rock band Autograph. Geopolitics, red tape, and suspicion stood in the way, but if they couldn't be there in person, we would bring them live via satellite.

That's where the magic of technology came in. We rolled in with our transportable uplink earth station, a towering, high-tech behemoth more suited to a missile launch than a music festival. Its massive parabolic dish, banks of electronics, and tangled web of cables were intimidating to most, but to us, it was a bridge to the impossible. Through that station, we could transmit live signals anywhere in the world. And this time, we were about to create a two-way international link, connecting San Bernardino directly with Moscow.

As the desert sun scorched the festival grounds, we carved our way through a sea of sunburned, dancing fans, escorted by police to keep everyone clear of the high-powered broadcast. Every knob, every cable, and every flick of a switch mattered. A single misstep could sever the link, and with it, the chance to make history. Then, as if by sheer engineering miracle and nerve, the connection stabilized. For the first time ever, an American festival and a Soviet stage were linked live, sending not just music, but a message of unity across one of the planet's deepest divides.

The massive screens flickered to life. Autograph appeared. For a heartbeat, there was silence. Then the crowd erupted. Thousands of hands shot into the air. Electrified energy rippled through the audience, flowing across the waves to Moscow. Autograph's lead singer took a breath and launched into the first song. Music became a statement, a bridge, a declaration that for one brief, glorious moment, politics and geography were irrelevant. Fans in California cheered, swayed, and even sang along, despite not understanding a single word of Russian.

Meanwhile, back in Moscow, an audience of Soviet citizens, many seeing a live American broadcast for the first time, watched in awe. For them, it wasn't just a concert; it was a raw, uncensored glimpse into a world they had only heard about through state-filtered headlines and whispers from the West.

Autograph's performance became the stuff of legend. It was a defining moment not only in their career but also in the broader history of cultural diplomacy, where music transcended borders and bridged the divide between two superpowers via satellite communications.

As for us, the team who made the connection possible, we packed up our equipment, navigated through a few more stumbling festivalgoers, and left knowing that, for one night, we had used technology not for war, but for unity.

A Three-Day Music Odyssey

Over three blistering days, the US Festival became a genre-defying odyssey. Van Halen roared with heart-shaking power, The Police delivered razor-sharp rhythms, and Fleetwood Mac's lush harmonies floated over the desert air. Tom Petty and the Heartbreakers unleashed raw, rebellious energy, while Talking Heads dazzled with art-rock brilliance. The Ramones, Santana, the B-52s, and countless others kept the pulse pounding. Every note resonated deep within the audience, transforming the festival grounds into hallowed ground for music lovers.

Through it all, the transportable uplink earth station hummed tirelessly, transmitting and receiving signals, linking audiences across continents. Wozniak's vision had been realized: music, satellites, and daring ingenuity had built a bridge over one of history's deepest divides. For those who danced in the desert and those who watched in Moscow, the US Festival wasn't just a concert; it was history in motion, a three-day music odyssey that fused culture, technology, and human connection in a way the world had never seen before.

LIVE AID: THE BIGGEST STAGE ON EARTH

The date was July 13, 1985. Live Aid shattered borders as technology spanned continents and united the world in a single moment of shared purpose.

The world was suffocating under the weight of a humanitarian nightmare. In Ethiopia, famine had already claimed nearly a million lives, and the horror showed no sign of slowing. Haunting images of skeletal children, cracked earth, and makeshift refugee camps flooded television screens, searing themselves into the global conscience. For Bob Geldof of the Boomtown Rats and Midge Ure of Ultravox, silence wasn't an option. Action had to be louder than the suffering.

At precisely noon in London, the first chords of Live Aid thundered through Wembley Stadium. Seventy-two thousand fans roared as the music ignited, while millions more watched from living rooms, bars, and public squares around the world. Across the Atlantic, the second act waited at Philadelphia's JFK Stadium, where the American half of this unprecedented global relay would unfold.

In all, more than sixty acts performed on both sides of the ocean, joined by additional performances beamed in from Australia, Japan, Germany, Yugoslavia, the Soviet Union, and Canada. However, connecting these stages was a feat no one had dared attempt before, let alone connect them live to a global audience.

The only way to pull it off? A sprawling, meticulously coordinated web of satellites. More than a dozen satellites linked continents and stitched together sixteen straight hours of music into a seamless, real-time global transmission. High above the Earth, geostationary satellites became the invisible threads of a transmission tapestry—rock 'n' roll, woven through space.

Live Aid proved that when people are inspired and technology is harnessed for good, the world can come together in breathtaking, unstoppable harmony. It was a landmark not only for humanitarian aid, but also for the power of satellite broadcasting itself—a triumph of space-age communication, demonstrating that it could do more than entertain; it could unite, uplift, and make history.

Thanks to those satellites, Live Aid became more than music. It became a global movement, broadcast through the heavens, fueled by song and forever etched in memory. The event raised more than $125 million for famine relief, and an estimated 1.5 to 2 billion people across 150 countries watched the moment unfold, all thanks to the largest, most ambitious live satellite broadcast the world had ever seen.

CHAPTER 10

SECOND-GENERATION SATELLITES AND TWO-DEGREE SPACING

SECOND-GENERATION DOMESTIC SATELLITES (1981-1985)

A quiet revolution was unfolding above our heads, one that would forever transform television. Second-generation domestic satellites were the engines behind this transformation, turning a limited and cumbersome industry into a broadcasting powerhouse. These orbiting giants, equipped with up to twenty-four transponders, brought more channels into American homes than ever before. Larger solar arrays and stronger 7.5-watt amplifiers delivered crystal-clear, static-free signals, making high-quality television a reality for millions.

The real breakthrough came with the introduction of Ku-band frequencies. Unlike the older C-band, Ku-band allowed for smaller, more affordable ground antennas. This innovation put high-quality satellite transmission within reach of cable companies, independent stations, and niche broadcasters, capabilities once reserved for the major networks. The impact was immediate. Cable exploded with specialized channels— twenty-four-hour news, round-the-clock sports, and live global events became the norm. Real-time reporting no longer felt extraordinary; it became standard. Suddenly, the world felt smaller, closer, and more alive.

By the early 1980s, North American communications had entered a new era. Second-generation satellites rose into geosynchronous orbit, delivering stronger signals, greater capacity, and continent-wide coverage. They powered cable networks, corporate feeds, and emerging direct-to-home services, not just expanding capacity, but tearing down distances. Orbit became a blazing superhighway, slamming real-time communication across North America into overdrive.

Second-generation satellites didn't merely improve television; they reinvented it. They transformed a rigid, local medium into a boundless, interconnected universe of information and entertainment, redefining how data moved, how news traveled, and how the continent stayed connected. These machines were more than technology; they were the heart of a communications revolution.

C-Band vs. Ku-Band

High above the Earth, suspended in the silent black of space, America's first domestic satellites drifted like watchful sentinels. Each carried a single, unwavering C-band signal, a thin, invisible thread that would shrink the world, bind the nation, and carry life itself into living rooms across the country.

By the early 1980s, a new generation of satellites arrived, pairing the steadfast C-band with the daring, nimble Ku-band. Antennas shrank. Signals sharpened. Television could now chase the story anywhere, from hurricane-ravaged towns to stadiums and city streets. Cameras rolled, satellites hummed, and events once delayed by hours or days leapt instantly into living rooms, crystal clear, alive.

C-band carried the weight. Ku-band gave it wings. Together, they didn't just transmit television, they brought it to life. Every goal, speech, and disaster could be shared with millions at once. Television had stopped imitating life; it had become life itself, in real time.

The Broadcast Networks' Embrace of Satellite Technology

In the late 1970s and early 1980s, America's broadcast networks stood at a crossroads, facing a choice between the tried-and-true and the daring new frontier. ABC and CBS clung to the familiar strength of C-band. Its broad, steady beams shrugged off storms and rain fade, delivering reliable coast-to-coast programming. For these networks, C-band wasn't just technology; it was assurance. Viewers could tune in from Maine to California and know the signal would hold, uninterrupted, a thread of consistency in a rapidly evolving broadcast landscape.

NBC, meanwhile, was chasing the future. It embraced Ku-band, the high-frequency upstart with smaller antennas, easier installation, and lower costs. Ku-band's narrow, focused beams allowed NBC to squeeze more content into the same spectrum and, crucially, put reporters on the ground anywhere, anytime. From city streets to rural communities, satellite news gathering brought breaking events live into living rooms, giving viewers a window to the world they'd never had before. Television was no longer chained to the studio; it followed the story out into the real world.

PBS had long been tethered to terrestrial microwave networks that struggled to reach every local affiliate. But with the rise of C-band satellite distribution, the public broadcaster gained new power: the ability to beam educational programs, cultural events, and live national specials directly into every corner of the country. Communities once isolated were now connected, and satellite technology had transformed from a technical resource into a lifeline.

In just a few short years, the networks' embrace of satellites reshaped television itself. Reliability, speed, reach, and mobility converged in orbit, giving the medium a new pulse, and the nation a shared window into its stories, unfolding live, in real time.

TWO-DEGREE ORBITAL SPACING

By the early 1980s, the rapid expansion of the U.S. domestic satellite industry had created a pressing problem: orbital congestion. With operators like Western Union *Westar*, RCA *Satcom*, Hughes *Galaxy*, AT&T *Telstar*, and GTE *Spacenet* launching satellites at a rapid pace, the FCC faced a critical challenge: how to allocate crowded geostationary slots efficiently while minimizing interference.

Previously, satellites in geostationary orbit had been spaced at least four degrees apart, a system that worked when only a handful were in operation. But surging demand for C-band and Ku-band transponders, driven by television, telecommunications, and corporate data networks, quickly filled the available slots. Between 1981 and 1985, the U.S. industry saw an unprecedented wave of launches, further intensifying the bottleneck.

To address this, the FCC adopted a two-degree spacing rule, effectively doubling the number of available orbital slots. This shift was made possible by advances in satellite technology, including improved antenna designs that reduced interference, enhanced frequency coordination to prevent overlap, and stricter power-control protocols to keep signals contained.

The results were transformative. More slots meant more operators, greater competition, and lower costs for broadcasters, telecom providers, and end users. The policy fueled the explosive growth of cable television, corporate data networks, and government communications. It also laid the foundation for the next generation of satellite services, from direct broadcast satellite (DBS) systems like DirecTV and Dish Network to very small aperture terminal (VSAT) networks, enabling businesses to establish secure, reliable private satellite connections.

By doubling orbital capacity, the FCC not only solved a critical logistical challenge but also ushered in a new era of satellite innovation, forever transforming how the world communicates and receives information.

Shaping a Lasting Legacy

Before the two-degree spacing policy, only eight to ten commercial geostationary satellites operated within the North American Geostationary Orbital Arc. By 1985, that number had more than doubled, a testament to the policy's immediate impact.

By the mid-1980s, this more efficient use of orbit fueled breakthroughs across industries. Expanded bandwidth made direct-to-home satellite television a reality, delivering high-quality entertainment to households across the nation. Corporate satellite networks also thrived, enabling businesses to transmit data and operate on a truly global scale.

The FCC's two-degree spacing rule stands as one of its most visionary decisions, transforming satellite communications and shaping the industry for decades. By maximizing the potential of geostationary orbit, it sparked innovation, expanded connectivity, and ensured that satellite services could keep pace with an ever-growing demand for them.

BEAMED, BOOKED, AND BUSTED

Navigating the satellite industry in the early 1980s meant grappling not only with fast-moving technology, but also with the maze of federal oversight wrapped around it. By late 1983, second-generation satellites were rising with impressive new capabilities, and the FCC's two-degree spacing mandate was redefining how we engineered and delivered every transmission. My days blurred into a cycle of filings, engineering reports, and calls with Washington—an intricate dance with regulators who could approve, delay, or derail multimillion-dollar operations with a single decision.

What I didn't realize was that my next encounter with the federal government would have nothing to do with satellites or filings. Just when I thought I understood Washington's reach and believed the FCC was the only agency capable of tightening the screws, two men in dark suits walked into my office unannounced. They weren't investigators or auditors. They were Secret Service agents, and they had come with pointed questions about a gaming tournament on a Native American reservation for which I was providing satellite transmission services.

In that moment, I learned a truth that every satellite professional eventually discovers: the federal government has many faces, many interests, and many doors—and you never know which one will swing open or tap you on the shoulder—or why—until the moment it happens.

It was late November, just before Thanksgiving, and everyone was in high spirits when two men in tailored suits appeared in my reception area. No flashing badges, no dramatic entrance, just a quiet authority that made people sit up a little straighter.

My normally unflappable secretary—who could juggle CEOs and news anchors without smudging her lipstick—leaned in with wide eyes. "They're from the Secret Service," she whispered. "They want to see you."

Moments later, the agents stepped into my office. Skipping pleasantries, one reached into his coat, pulled out a photo, and slid it across my desk like a poker chip.

"Do you recognize this man?"

I leaned in. The face was familiar. "Yes. He was here a few weeks ago."

He had pitched a quirky but harmless idea: a live bingo tournament broadcast from a Native American reservation to multiple locations across the Southwest. We signed a contract, the payment cleared, and I moved on. Business as usual—or so I thought.

"This man," one agent said quietly, "is a known con artist. He's been under surveillance."

The second agent added, "The governor requests that you withdraw your support for this event."

I hesitated for a brief moment, gathering my thoughts. "I understand," I replied, "but I'm bound by a signed contract and I've already received full payment. If the governor provides a written release from all liability, I'll cooperate."

They exchanged a quick glance. "We can't do that."

Without further comment, they stood, thanked me, and walked out, leaving more questions than answers.

Their warning gnawed at me. Only later, after some digging, did the truth come into focus: the slight delay in satellite transmission—barely a quarter of a second—was enough for organizers to manipulate the game in real time, quietly altering the numbers being called to rig the outcome. What had seemed like a routine broadcast setup was in fact the backbone of a high-tech scam.

Despite the strange circumstances, the event proceeded as scheduled. The broadcast went off without a hitch—the satellite feed was flawless—and I began to think the ordeal was behind me. Then my phone rang.

It wasn't the governor; it was the uplink operator. His voice carried an uneasy edge. "Law enforcement just commandeered the tournament organizer's vehicle," he said.

I paused to absorb the news. "Are you okay?"

"I'm fine," he assured me. "They're telling me to carry on as planned."

As the story unfolded, I learned what happened next: the moment my client stepped off the reservation and onto U.S. soil, the federal agents were already in position. Weeks of tracking, movements mapped, every square on the board quietly marked. He didn't even sense the trap closing in. The cuffs snapped on before he could react, swift and precise, a move he never saw coming.

Every square, every number, every play he thought he controlled— they had already been claimed. The final call had been made. The board was theirs. The prize secured.

He didn't realize the Feds had his number all along.

BINGO!

Game Over.

The moon landing marked a global turning point for humanity, a defining moment when millions didn't just hear about history, they *watched it unfold live via satellite*.

For the first time, the world watched a live television broadcast from China as President Nixon met with the nation's communist leaders—a historic moment that showcased both a breakthrough in diplomacy and the power of satellite technology in real time.

The signing of the European Broadcasting Union contract in Lake Geneva secured the international transmission rights for the 1984 Los Angeles Summer Olympic Games, a pivotal milestone that helped shape one of the most-watched global events in history. *Bob Patterson seated fourth from the left in the dark suit.*

Mr. and Mrs. Rupert Murdoch
cordially invite

Robert & Nancy Patterson

To a Gala International Reception
to celebrate the occasion of
the 1984 Los Angeles Olympic Games
on Wednesday, 1st August,
5.30-8.00 p.m.
in the Ballroom
Beverly Wilshire Hotel,
Los Angeles, California.

Invitation to Rupert Murdoch's Network 10 Australia gala, celebrating the network's record-breaking broadcast of the 1984 Los Angeles Olympic Games. Murdoch, the Australian-born media mogul, had by then built one of the world's most influential global media empires.

The artistic rendering of the Olympic torchbearer on the gala invitation from Mr. and Mrs. Rupert Murdoch, commemorating Network 10 Australia's successful broadcast of the 1984 Los Angeles Olympic Games, symbolizes the triumph, energy, and global impact of the Games.

Westar I, launched in 1974 by Hughes Aircraft Company, became America's first domestic communications satellite. A technological milestone, it ushered in a new era of broadcasting and transformed how the nation experienced sports, news, and entertainment.

Through innovative use of a compact 2.4-meter antenna, SPACECON-NECTION's affiliate, Mobile Satellite Connection, was able to transmit analog, digital, SD, and HDTV signals, setting a new standard for trans-portable uplink services.

The Frost/Nixon interviews marked a landmark in broadcast journalism, drawing record audiences and becoming the most-watched political interview in television history. *Bob Patterson, right, in the control room during the premiere airing on May 4, 1977.*

Telstar 402R was launched from the Guiana Space Centre in Kourou, French Guiana, under Arianespace's operation, marking a triumphant return for AT&T while reaffirming its leadership role in the satellite industry.

Promoted by Don King Productions, the June 11, 1982, WBA Championship bout between Gerry Cooney and Larry Holmes became a knockout success, breaking records as the largest pay-per-view event of its time.

The 1984 Los Angeles Summer Olympics shattered records, reaching over a billion viewers worldwide and setting a new standard for flawless execution, global television coverage, and commercial triumph.

On October 29, 1988, John Glenn's return to space aboard the Space Shuttle Discovery became a spectacular milestone in space history, brought to life like never before through NASA's HDTV coverage.

As the Space Shuttle blasted into orbit, it opened a new chapter in spaceflight, transforming technological innovation into reality, pioneering regular missions, deploying satellites into orbit, and returning safely to Earth to be flown again.

Suspended in the vastness of space, astronauts accomplished a historic first, retrieving a satellite and carefully stowing it in the Space Shuttle's cargo bay, showcasing human ingenuity and technological mastery. Photo courtesy of NASA.

At Hughes Aircraft Company's Space and Communications facility in El Segundo, two satellites returned from space are carefully refurbished, a testament to an unprecedented astronaut achievement and the pioneering work of Hughes' engineers. *Bob Patterson kneels in the second row, third from the left.*

On May 23, 2002, during the dedication of the Marlin Fitzwater Center for Communication, a memorable photo captured Marlin Fitzwater, Nancy Patterson, President George H.W. Bush, and Bob Patterson, commemorating a distinguished moment in honoring Fitzwater's legacy at Franklin Pierce University.

Marlin Fitzwater stands before the Fitzwater Center for Communication at Franklin Pierce University, the building bearing his name in tribute to his legacy.

Standing between Presidents Reagan and Bush, Marlin Fitzwater is celebrated for his remarkable tenure as White House Press Secretary, serving a pivotal role in communicating the major geopolitical events unfolding around the world.

During a Franklin Pierce University gala honoring Marlin Fitzwater, Nancy and Bob Patterson, share a memorable photo with President George H.W. Bush and First Lady Barbara Bush.

Bob Patterson shares an unforgettable moment with golf legend Arnold Palmer at The Golf Channel Invitational, an encounter made possible by the pivotal role Bob's company played in transmitting The Golf Channel's broadcast events.

CHAPTER 11

ARE YOU READY TO SCRAMBLE?

THE MAIN EVENT

The sound system roared to life, the announcer's voice booming across the arena like a cannon blast: "Live from the legendary Caesars Palace, Las Vegas, under the stars of the Nevada desert, the moment has arrived. Two men; one destiny. This is the fight of the decade!"

The tension in the arena was a living, breathing, suffocating force as 30,000 fans erupted with anticipation. Hollywood's biggest names filled the front row, their faces a mixture of excitement and awe, knowing they were about to witness something unforgettable.

"Good evening, ladies and gentlemen, and welcome to what promises to be one of the most legendary nights in boxing history. The atmosphere is electric. The air is thick with anticipation as we prepare for the WBC World Heavyweight Championship between the undefeated challenger, Gerry Cooney, and the reigning, defending champion, Larry "The Easton Assassin" Holmes!

The crowd exploded, and the atmosphere shifted into an uncontainable frenzy.

"Making his way to the ring: the undefeated knockout artist, fighting out of Huntington, Long Island, New York, Gerry "The Gentleman" Cooney.

At six feet six inches, Cooney was a towering presence draped in white and green. His entrance was commanding, a fusion of grace and raw power. The crowd's reaction crashed over him like a tidal wave of emotions, cheers mingling with boos, admiration clashing with rivalry, a turbulent storm about to begin.

The announcer let the energy simmer for a heartbeat, building the tension before continuing.

"And now, for the defending WBC Heavyweight Champion of the World, fighting out of Easton, Pennsylvania, Larry "The Easton Assassin" Holmes!"

Holmes emerged with the kind of swagger only a seasoned champion could possess. His gold-trimmed robe shimmered under the lights, radiating authority. With a jab that had dismantled legends, Holmes exuded calm, focused power, the kind forged by time and unshakable belief.

The crowd's roar swelled to a deafening fever pitch. In the center of the ring, the unmistakable figure of Michael Buffer stepped forward. His grin flashed in the dimming lights as the crowd hushed, breathless for the moment they all craved.

A microphone descended from above. Buffer let the tension simmer, soaking in the silence before unleashing the words that would ignite the night:

"And now . . . for the thousands in attendance . . . and the millions watching around the world, "LET'S GET READY TO RUMBLE!"

The arena erupted, cheers, screams, and roars crashing against the ring like a hurricane slamming into the shore. The moment had arrived, but what the roaring masses couldn't see, what no one could even begin to imagine, was that a different kind of battle had already ignited. Far from the blinding lights and the glittering ringside seats, behind a cold steel door, a different kind of war raged.

BEHIND THE STEEL DOOR

It started months before the "Clash of the Titans," with an invitation to a production meeting at Don King's office in Manhattan.

This wasn't the sort of place you happened upon. It was the kind of street that didn't welcome visitors, and those who did come rarely stayed.

Off 6th Avenue, wedged between two anonymous storefronts, stood a steel door. It wasn't an entrance. It was a dare. I loitered in front of the steel door, trying to find an address number when he appeared.

A refrigerator disguised in a suit. Six-five, maybe six-six. Neck like a tree trunk. He had hands that looked like they could bend rebar for fun.

"What are you looking for?" he growled, his voice pure gravel.

"Don King's production meeting," I answered, steady but cautious, the tone you use when you're hoping not to get flattened.

No smile, no handshake, just two sharp raps from the giant's knuckles against the steel. A tiny hatch, framed by rusted iron, slid open, revealing a pair of cold, suspicious eyes peering through the slot.

Then came the harsh click of deadbolts, and the steel door creaked open like the awakening of a long-forgotten crypt. For a moment, it felt like I'd stumbled into a Prohibition-era speakeasy, except there was no smooth jazz and no velvet drapes—just the raw, suffocating tension of a noir gangster film.

The air changed, colder, heavier, thick beneath a shroud of gloom. Inside, a narrow, battered elevator waited like a trap. Beside it stood a man whose rumpled jacket barely concealed the pistol tucked into his waistband.

I hesitated. Was this a meeting or a one-way ticket to the bottom of the Hudson River?

We rode up in silence. No small talk, just the low hum of danger closing in. At the top, a grim conference room waited, its walls seeming to lean in, as if eavesdropping. Around the battered table sat men whose eyes had seen too much and spoken too little.

The moment I stepped inside, a cold shiver ran down my spine, a quiet warning that what was coming would be menacing, unwelcome, reckless. A question I didn't want to face, yet one I knew was inevitable, hung in the air as I slowly lowered myself into the chair.

At the head of the table sat Jack, the production boss. His only greeting was a curt nod: *You're here. For now.*

The meeting trudged through the usual rundown, production schedules and site plans, until the air shifted. Then it came: my turn. The real reason I was there: the satellite transmission for the fight the world couldn't stop talking about, Holmes versus Cooney.

The plan was tight: independent tech teams, TVRO downlinks, a sleek, flawless live pay-per-view feed stretching coast to coast. Everything lined up like clockwork, until Jack leaned in, voice low, and dropped the bomb.

"What about scrambling the signal?"

The room went deadly quiet. The words hit like a shotgun blast. Scrambling, one word, loaded with nightmare scenarios: signal theft, piracy, freeloaders hijacking the fight without paying a dime.

No one moved. No one breathed. All eyes turned to me, waiting for the crack, the collapse, the desperate backpedal. Jack's gaze sharpened, cold as steel.

"Too risky," I said. "If even one decoder fails, that site goes dark. You lose the feed. The fight's gone."

Jack slammed his fist on the table. His voice dropped to a growl. "Don King doesn't want anyone watching for free."

I chose my words carefully, knowing I was walking a razor's edge. "The scrambling system isn't ready for prime time. We're gambling millions on tech that could fail without warning."

Jack's gaze turned to ice. No room for debate now. "Then we got a problem," he said slowly. "Might have to pull the plug on the whole contract."

The room froze. His words hung heavy in the air. All eyes locked on me, waiting for the crack, the plea, and the retreat.

I swallowed hard. "We can't risk it. Not on this fight."

The silence stretched, sharp enough to slice through steel. Then Jack erupted. "THEN GET OUT!"

His voice ripped through the room like a thunderclap, no arguments, no room for second-guessing, just cold, unyielding steel in every word.

I stood up, my chair scraping loudly against the worn floor. The room seemed to tilt as I pushed past the stunned faces, each step heavier than the last.

The hallway swallowed me whole. Behind me, the door slammed shut; the brutal punctuation on the man who dared defy Don King's iron will.

The deal was dead. The biggest payday of my career vanished in an instant. All that lingered was the echo of Jack's roar and the cold, empty finality of a door that would never open again.

I caught a late flight back to Los Angeles, pride bruised but unbroken. I had just turned my back on fame, fortune, and a fast track to power, but I hadn't sold my soul.

In a world ruled by backroom deals and shady shortcuts, I had chosen principle over profit.

THE SATELLITE STING

Days after the steel door slammed shut in New York, the phone rang. It was Jack. The bravado, the bark, and the swagger were gone. Now his voice carried the weight of a fighter pinned against the ropes, searching desperately for a last opening.

"Bob," he said, almost a whisper, "you change your mind about scrambling?"

A long pause. Calm, steady, unflinching: "No."

But Jack wasn't ready to throw in the towel, not yet. "Listen," he pressed, desperation creeping back into his tone, "your company's the best. You know it. We know it. But Don . . . Don's dug in. He won't let this air unless we lock it down. Every pair of eyes has to pay."

This was no ordinary network. Nearly 215 transportable downlink sites stretched across the nation, an undertaking never attempted before. If coordinating a massive 123-site televangelist broadcast was like juggling dozens of spinning plates in a wind tunnel, managing nearly double that number of TVRO sites was like piloting a spacecraft through a meteor storm. One misstep, one glitch, and the entire network could tumble into the void.

Jack knew it. An event of this magnitude could collapse in an instant if entrusted to anyone untested under such pressure. Every satellite, every signal, and every timing window was a fragile thread holding the broadcast together, and if one snapped, the whole operation could unravel. Yet here he was, cornered, forced to bend to Don King's demands, the weight of the nation's gaze pressing down on him. His mind raced, calculating every risk, imagining every way it could all go wrong. One wrong call, one missed cue, and the operation would implode, leaving him to face consequences too vast to fathom.

Then, almost like a man admitting defeat, Jack's voice softened. "Is there any way to make this work?"

On the other end, silence settled again—the kind of silence that speaks louder than words ever could. Behind it, gears turned slowly, methodically, driven by years of field-tested knowledge and battlefield experience.

Finally, the stillness broke. "Maybe," came the response.

Back then, analog satellite feeds were open season for signal pirates. With a backyard dish and a little know-how, anyone could hijack a live transmission as casually as turning on the evening news. Outsmarting them—that was the real challenge. That's when I conceived *The Satellite Sting*: two feeds, one real and one fake. A high-stakes bait-and-switch, not played in smoke-filled back rooms, but across the silent, black expanse of space. Our marks—the pirates—would sail straight into a trap any seasoned scallywag would envy.

"Alright," I told Jack. "Here's how the plan plays out."

"We'd seed the sky with a decoy feed, broadcasting from the far edge of the orbital arc with a bogus transponder number, a phony event slate, even reversed polarization—just enough to throw off the most hardened satellite thieves and backyard hackers. While they chased ghosts, the real feed stayed locked down tighter than Fort Knox, accessible only to a tiny inner circle of double-verified techs."

I leaned back slightly, a sly grin appear. "And the pirates?" I added, letting the question drip with playful menace. "All their clever tricks, all their scheming . . . their antennas would be aimed at nothing but an empty promise. Two hours of static. Not a signal, not a whisper—just a blank slate mocking every plan they thought would outsmart us."

STAY TUNED TO:

HOLMES VS. COONEY CHAMPIONSHIP FIGHT

FROM CAESARS PALACE

JUNE 11, 1982

It was a textbook sting—a flawless masterpiece of deception. *"We'd walk away with the payday," I said, "and they'd be left chasing shadows, bewildered, befuddled, and furious at a prize that wasn't there."*

Jack's voice lit up, sparking with excitement. "Brilliant! Let's do it!"

With those words, the fight roared back to life.

The plan was set. Now came the hardest part: the execution.

Jack sent over the master list of authorized venues. Phones lit up, most legitimate, some sniffing for pirate access. Every call was vetted—every venue, verified. The real coordinates were shared only with a trusted few. No leaks. No mistakes. No mercy.

Over 200 theaters, arenas, and stadiums across the country geared up for the live feed blitz. From dusty backroad movie houses to gleaming city stadiums, each site location had one mission: flawless transmission. No exceptions.

Technicians moved with military precision, double-checking set-ups and reconfirming connections. Test signals blazed through the night like rockets piercing the sky. Receivers were locked in with pinpoint accuracy of NASA ground control.

Nothing, absolutely nothing, was left to chance.

And then, the big moment. Under the neon glow of Caesars Palace, the wait ended.

It was electric. Across America, crowds roared. Big screens flickered to life. Beer sloshed. Fists pounded the tabletops. Voices rose in a single, thunderous wave of anticipation. And the signal? Perfect.

The atmosphere at Caesars was supercharged with excitement.

Referee Mills Lane brought Holmes and Cooney to the center ring, his voice slicing through the noise: "Alright, gentlemen, you know the rules. I want a clean fight. Protect yourself at all times. Obey my commands. Touch gloves now . . . and let's come out fighting."

They touched gloves and then stepped back to their corners.

And then, the first bell exploded through the air. Months of meticulous planning and technical wizardry surge to life.

Larry Holmes was a fortress: endurance, intelligence, and tactical brilliance personified. Gerry Cooney, all raw power and fury, came crashing in like a thunderstorm.

Holmes absorbed the blows, danced along the ropes, and struck back with sniper-like accuracy. In the thirteenth round, he delivered the decisive blow, a TKO victory; a masterclass in boxing under the brightest lights.

But behind the curtain, the real triumph wasn't just in the ring, it was in the transmission. From coast to coast, city to town, it was a flawless broadcast, a living, breathing masterpiece. Every downlink site hit its mark. Every frame, every moment, delivered in real time and without a single glitch.

The nation tuned in for Holmes vs. Cooney, a knockout success, while the public remained unaware of the storm of chaos that brewed behind the scenes.

And the pirates and freeloaders who expected a free ride were met with nothing but a frozen slate, locked out, left stewing in silence until the evening news tossed them scraps. Meanwhile, Holmes vs. Cooney didn't just break records; it obliterated them, as Don King Productions hauled in $50 million on the back of the largest satellite downlink network ever assembled.

THE ENCRYPTION SYSTEM WORKED TOO WELL

Fights like Holmes vs. Cooney come along maybe once in a decade, explosive, unforgettable, the kind of spectacle that burns itself into memory. For one electric night, we weren't just broadcasting a fight; we were holding history in our hands. Every eye in the nation was locked on the screen; every heartbeat seemed to rise and fall with the clash in the ring.

And then . . . silence. The lights dimmed, the roar dissolved, and reality crept back in. Just like that, we were back to the grind: video teleconferences, religious broadcasters, and concert promoters.

I was still riding the adrenaline rush from that night when the phone rang. On the line was one of our marquee Silicon Valley clients, the kind of company known for always staying three steps ahead of the curve. But their message hit like a right hook to the ribs: they were leaving us. They'd been seduced by promises of next-generation encryption, dazzled by shiny technology, and convinced that another satellite provider was better equipped to handle their upcoming video conference.

In satellite broadcasting, encryption is the lock on the vault. The feed gets scrambled by an encoder, and only a decoder armed with the right algorithm, the secret key, can set it free. On paper, it was airtight, unbreakable. In practice? It was often fragile, temperamental, and not nearly as bulletproof as advertised.

The executives brimmed with confidence as the encrypted teleconference approached. "It will be seamless, secure, and unbreakable," they assured everyone. They even boasted that not a single hacker, no matter how resourceful, could intercept their broadcast.

On the big day, fifteen sites across the country were primed and ready. The countdown began. The feed went live. Anticipation filled the air, followed by despair. Only three of the fifteen sites successfully received the signal. The rest? Blank screens. A dozen conference rooms full of frustrated executives, staring in stunned silence at a whole lot of nothing.

A few days later, curiosity got the better of me. I called the president of the satellite firm and asked the obvious question: "Was it the encryption system that failed?"

I braced for a parade of excuses: bad software, rogue technicians, misconfigured hardware. Instead, he chuckled and said: "Actually, the encryption system worked too well. We just couldn't unscramble the signal."

His lighthearted tone offered a flicker of levity, but beneath it lay the brutal truth: the system designed to protect the broadcast had betrayed them. Instead, the very technology meant to safeguard the teleconference had backfired spectacularly, locking out its own audience.

As I hung up, the what-ifs came fast and furious. What if Holmes vs. Cooney—the fight of the decade—had gone dark? No picture. No sound. Dead screens across America.

Pressure screamed to chase new tech. We didn't. We trusted our gut, leaned on experience, and remembered the rule: never gamble with history.

The fight aired flawlessly—bright, sharp, alive. Every punch landed. Every roar rang out.

By rejecting the hype, we dodged the TKO (technical knockout)— and emerged victorious.

CHAPTER 12

1984

THE YEAR THE FUTURE ARRIVED

1984 wasn't just another year on the calendar—it was a pivotal moment, a time when the world quietly began its shift into the digital age. It was a period charged with possibility, when technology was accelerating faster than anyone could have imagined. At the center of this transformation stood two forces destined to change everything: satellite communications and the personal computer.

I found myself swept up in that wave of innovation, striving to keep pace with a rapidly changing world moving at light speed.

My year didn't begin behind a desk—it started under the Florida sun at Super Bowl XVIII, not as a spectator, but as the person delivering CBS's live broadcast to the nation. Even as the game loomed, preparations for another global spectacle—the 1984 Summer Olympics—were already underway.

However, on January 22, 1984, my attention was focused on the "Big Game." From Tampa Stadium, I oversaw the live satellite transmission to CBS in New York, where commercials were inserted before the signal reached millions. Every cheer, every replay, every yard traveled along invisible highways of space, connecting the stadium to living rooms across America.

While the Raiders and Redskins battled for glory on the field, another far more consequential contest was quietly being waged in the corporate world. This was the night Apple fired the first shot of a technological revolution.

Apple had built a revolutionary machine, the Macintosh, but it wasn't the computer itself that detonated across America that night. It was the commercial, not just any commercial, but the one that would unleash a cultural shockwave.

As the third quarter unfolded, CBS launched a 60-second commercial unlike anything television had ever seen. Apple's "1984" ad erupted across screens nationwide—a dystopian fever dream brought to life by Ridley Scott. Millions of viewers, expecting another beer or car commercial, were instead thrust into a gray, oppressive world where a lone heroine hurled a sledgehammer through the looming face of conformity.

The screen shattered— and the past gave way to the future.

It was a declaration, a rebellion, an awakening.

In that single moment, Apple didn't just unveil the Macintosh, they announced the dawn of the personal computer.

And fittingly, it was a satellite that carried that declaration across the nation. While Apple promised personal empowerment, satellites were already reshaping the global stage. They collapsed geography, erased the limits of time, and transformed local events into shared experiences. Political summits, hostage crises, Olympic ceremonies, championship games—no longer isolated moments. Satellites made them communal, immediate, universal.

Where one technology placed power in the hands of the individual, the other connected those individuals to the entire planet.

Their convergence wasn't accidental. It was inevitable.

In retrospect, the Super Bowl feed and Apple's "1984" ad were more than broadcast milestones. They marked the beginning of an interconnected era, a world where the boundaries of time and distance no longer constrained the flow of knowledge, ideas, and culture.

You have just crossed into another dimension—**Welcome to the Information Age.**

BEAMING DEMOCRACY

The roar of Super Bowl XVIII had barely faded when the next high-stakes mission came into view. This time, it wasn't about football—it was about history itself.

Overseeing the Super Bowl while preparing for the 1984 Summer Olympics would have been enough for most people. But I wasn't ready to stop there. My sights were set on something bigger, riskier, almost unthinkable: convincing the European Broadcasting Union's thirty-five-nation alliance to trust us with both National Political Conventions.

These weren't routine broadcasts. They were democracy's most volatile theater, where politics performed like drama and diplomacy played out under unforgiving lights. In an election year, every satellite beam, every camera shot, every flicker across the screen carried history itself.

And when the final decision was made, we were once again entrusted with delivering satellite services on behalf of thirty-five European nations.

First up: the Democratic National Convention, set to transform San Francisco's Moscone Center into a stage of national drama from July 16 to 19. A little more than a month later, the spotlight would swing to Dallas, where Republicans would seize the stage at the Convention Center from August 20 to 23. Sandwiched between these political heavyweights? The 1984 Summer Olympics was a global broadcast marathon and a logistical behemoth in its own right.

By the time we entered the fray, ABC and NBC had already locked down most prime satellite slots for the Democrats. What remained were scraps, thin slices of orbital real estate dismissed by the giants as unusable.

But we weren't empty-handed.

We held a wild card: exclusive transponders and a fleet of transportable uplink trucks, nimble enough to reach places the networks couldn't. Technically, we had the firepower; what we lacked was time.

And looming over it all was the shadow we could never escape: frequency coordination. A relentless maze of regulations, filings, and red tape. One misstep—one mistuned number, one signature gone missing—and the mission would implode before it even launched.

THE ELECTROMAGNETIC CHESS GAME

Coordinating satellite and terrestrial frequencies wasn't bureaucracy, it was warfare on an invisible front. In media hotspots like San Francisco, Los Angeles, and Dallas, a single miscalculation could turn a pristine live broadcast into a static-filled disaster. The FCC wasn't a bystander; it was the unflinching gatekeeper of the airwaves. Before a single uplink dish could fire skyward, the Commission demanded full-scale frequency coordination. One misaligned signal could obliterate your broadcast—or someone else's.

The real minefield was the C-band, shared by satellite backbones and local microwave links. Every move was blindfolded chess against invisible opponents. Case in point: the Democratic National Convention at Moscone Center. Our first choice—parking an uplink truck outside—was dead on arrival. The spectrum was saturated. Broadcasting there would have been like trying to light a match in a hurricane.

We needed a miracle. After site surveys and spectrum scans, we found it: El Cerrito, a quiet suburb across the Bay. Far enough to escape radio frequency chaos, close enough for a flawless microwave line from the convention floor.

The plan was audacious. Temporary microwave relay dishes captured the live feed at Moscone and hurled it across the water to three transportable uplink stations in El Cerrito. From there, signals rocketed skyward, reaching the European Broadcasting Union, CBS Television Network, Taft Broadcasting, and affiliates nationwide.

Inside the convention hall, broadcasters focused on podiums, teleprompters, and carefully staged theater. Outside, an entirely different story unfolded: protesters, street performers, and eccentrics turned San Francisco's sidewalks into a live-wire portrait of democracy.

The breakout star? Not Walter Mondale, but Sister Boom Boom. Sequins glittered under the camera lights as this unorthodox nun of the Sisters of Perpetual Indulgence marched, twirled, and mocked the establishment with gleeful irreverence. For Europeans watching, it must have felt like stumbling into an alternate America, one that tossed aside solemnity in favor of satire, one where politics collided with performance art, and where drag nuns could steal the stage from presidential candidates.

Sister Boom Boom was not a sideshow; she was the show—the embodiment of America's eccentric freedom, a reminder that democracy doesn't live only in polished halls of power or scripted speeches. It lives in the streets, in voices and visions that refuse to be ignored. And on live television, it was unforgettable: authenticity wore sequins, a habit, and the name Sister Boom Boom.

THREE CITIES, THIRTY-NINE DAYS, AND SISTER BOOM BOOM

It was a summer that never paused, and neither did we. From the roar of Democrats in San Francisco, to the spectacle of the Los Angeles Olympics, and finally the Republican crescendo in Dallas, we raced against the clock, exhausted yet relentless.

When Walter Mondale finished his acceptance speech and the last balloon drifted across Moscone Center on July 19, most people exhaled. For political insiders, the week of caffeine, strategy, and spin was over. For broadcast networks, it was time to pack up. For us in satellite transmission, there was no "off-switch."

As confetti fell, we were already racing to Los Angeles. In eight days, the world would watch the XXIII Olympiad. While athletes chased gold, we orchestrated signal paths, managed uplinks, scanned spectrum, and dodged technical landmines. Barely catching our breath between one monumental broadcast and the next, we were already on the move.

Next stop: Dallas. The 1984 Republican National Convention waited, indifferent to our Olympic triumph. Trucks reloaded, crews reassigned, frequencies re-coordinated. We secured uplinks just beyond the convention center, rolled in transportable earth stations, set up temporary relays, and launched signals skyward.

On August 23, Reagan accepted his party's nomination. As his words echoed around the world, we finally exhaled. The sleepless nights, meticulous planning, and relentless pressure had all led to this. This wasn't just a broadcast; it was a symphony of logistics, technology, and sheer will.

In thirty-nine days, we leapt from San Francisco to Los Angeles to Dallas, powering through three of the decade's most grueling live events. Politics, global sports, and cutting-edge technology collided in a blur of live broadcasts and sleepless nights. Every moment demanded precision. Every broadcast carried the weight of history.

Yet, when the dust settled, it wasn't the speeches of politicians or the triumphs of athletes that lingered most vividly—it was the sequins, the sass, and the irreverent audacity, a drag nun who commandeered the spotlight like no one else: Sister Boom Boom.

Then, on November 6, 1984, it culminated with Reagan's landslide re-election. Headlines captured the politics, passion, and power plays, but another story was quietly written—in signals, satellites, and innovation. That summer, the sky became our conduit, and a daunting challenge became a defining moment in the history of satellite broadcasting.

Through it all, the American story unfolded instantly, coast to coast and around the globe. Satellite broadcasting had graduated from experimental novelty to indispensable infrastructure, shaping not only how we witnessed history but how, in 1984, the age of real-time, global communication had truly arrived—raw, unfiltered, and unstoppable.

CHAPTER 13

TRIUMPH AND TRAGEDY

A GLIMPSE OF NASA's FIRST SPACE SHUTTLE

In August 1979, I had the extraordinary privilege of being a special guest of Western Union at the Kennedy Space Center in Florida to witness the launch of *Westar III*. It was a breathtaking experience: the roar of ignition, the ground trembling beneath my feet, and a communications satellite soaring into orbit. But as unforgettable as that moment was, something even more remarkable awaited me just beyond the launchpad: a rare, behind-the-scenes VIP tour of NASA's revolutionary Space Shuttle prototype, *Enterprise*.

Enterprise was no ordinary spacecraft. As NASA's first orbiter, it never traveled to space, but its contributions were foundational. This test vehicle laid the groundwork for an entirely new era of space exploration. Unlike the single-use Apollo rockets before it, the shuttle was designed as a fully reusable vehicle, able to launch like a rocket, orbit like a spacecraft, and glide home like a plane. It was a radical leap forward; a machine built not just for one mission, but for many.

Originally intended to be named *Constitution*, the orbiter received its iconic name *Enterprise* after a deluge of letters from devoted *Star Trek* fans flooded the White House. President Gerald Ford responded, honoring not just a fictional starship but the hope and excitement it symbolized. That moment revealed the power of imagination and public passion. Space was no longer just for scientists and astronauts; it was becoming part of our collective dream.

NASA had bold ambitions for the Shuttle. It wasn't merely a vessel for astronauts; it was a versatile, orbital workhorse. Its cargo bay could launch satellites, carry scientific labs, and even return to space to retrieve or repair equipment already in orbit. For the first time, humanity had a spacecraft that could go to space, come home, and then go back again. It was, quite simply, a game-changer.

During my private tour, I was ushered into testing areas that few civilians had ever seen. There, among the hum of machinery and the careful choreography of engineers in white coats, I learned of a critical concern: the thermal protection tiles that shielded the Shuttle during reentry were failing to stay in place. The adhesive used to secure them was unreliable, and a complete reinstallation was underway.

At the time, the gravity of that issue hadn't fully registered. But in 1981, as Space Shuttle *Columbia* soared into orbit on its inaugural flight, the warning signs became clear. Over 300 tiles were lost during launch and reentry. Though the mission ended safely, it was a stark reminder of the razor-thin margins of error in space travel. NASA quickly improved its bonding processes and reinforced at-risk areas, but tile detachment would haunt the program for years.

Still, the dream was undeterred.

The Shuttle program represented a defining shift in our relationship with space. It promised more frequent missions, lower costs, and a future where orbit wasn't distant or rare; it was reachable. The optimism was palpable. Engineers, astronauts, and everyday people alike believed they were witnessing the beginning of something transformative. Space travel had stepped out of the realm of science fiction and into reality, achievable, reliable, and within humanity's grasp.

MISSION: IMPOSSIBLE

On February 3, 1984, Challenger roared into orbit on its fourth flight, carrying two satellites that embodied the future: Indonesia's *Palapa B2*, set to connect Southeast Asia, and America's *Westar VI*, a second-generation leap for U.S. communications. Their final push to geostationary orbit depended on the payload assist module.

It failed.

Instead of drifting into perfect synchronicity with Earth, both multimillion-dollar satellites were left stranded, too high to be easily retrieved and too low to function. It was the space-age equivalent of launching two luxury yachts, only to beach them in the middle of the desert.

The fallout was immediate. Owners panicked, insurers scrambled, and engineers scoured telemetry for answers. What should have been a showcase for commercial spaceflight had become a public disaster.

But NASA wasn't ready to give up.

What if they could bring them back?

It was radical. Satellites weren't designed for retrieval, but the Space Shuttle was different. Unlike rockets, it could haul cargo into orbit *and bring it home*. For the first time, a salvage mission in space seemed possible: intercept the marooned satellites, capture them, and return them to Earth.

On November 8, 1984, *Discovery* launched with a crew tasked with the unthinkable. Using its robotic arm, astronauts inched toward the helpless satellites, each maneuver a high-stakes ballet in zero gravity. One wrong move could doom the mission. Instead, the arm locked on, first *Westar VI*, then *Palapa B2*, guiding them gently into *Discovery's* bay.

When the Shuttle touched down at Kennedy, the impossible had been done. Two lost satellites, written off as failures, were back on Earth.

The story didn't end there. At Hughes Aircraft in California, engineers revived them. *Palapa B2* reemerged as *Palapa B2P*, relaunched in 1990 to finally deliver the service Indonesia had envisioned. *Westar VI*, reborn under Hong Kong's AsiaSat, found new life powering communications across the Asia-Pacific.

What began as a failure became a showcase of resilience, ingenuity, and the Shuttle program's unmatched versatility. NASA had proven that in space, as on Earth, defeat isn't final, it's an invitation to imagine what else might be possible.

THE DAY SPACE EXPLORATION CHANGED FOREVER

By 1985, America's Space Shuttle program was soaring, literally as well as figuratively. With four orbiters in rotation—*Columbia, Challenger, Discovery,* and the newest *Atlantis*—NASA had completed twenty-four missions at breakneck pace. The agency envisioned a future in which shuttles would launch like clockwork—perhaps even weekly—ushering in a permanent human presence in orbit. Civilian flights were on the horizon: teachers, journalists, even business executives were being considered for missions. It was an era electrified by possibility, where science fiction seemed to be folding into reality. Beneath the excitement, however, troubling signs were emerging.

As NASA pressed to meet its ambitious launch schedule, safety began to yield to speed. The culture had shifted. Confidence, once earned, had hardened into complacency. Engineers voiced concerns, particularly about the solid rocket boosters and the O-ring seals that could fail in extreme cold, but their warnings were often softened, overlooked, or lost in bureaucratic layers. Then came STS-51-L, the twenty-fifth mission of NASA's Space Shuttle program.

It was meant to be a defining moment, one that captured the hearts of millions. On board Space Shuttle *Challenger* was Christa McAuliffe, a high school teacher from New Hampshire and the first civilian chosen to fly into space. Her mission: to teach lessons from orbit and ignite the imagination of students across the country. America watched with pride. But nature had other plans.

The morning of January 28, 1986, dawned brutally cold, the kind that rarely visits Cape Canaveral. Temperatures dipped below freezing, well outside the shuttle's certified safety margins. Icicles clung to the launch tower. Engineers at Morton Thiokol, the booster manufacturer, raised urgent red flags about the integrity of the O-rings in such conditions, pleading for a delay. Time was tight, and the pressure to launch was immense. In the final hours, after a tense, controversial teleconference, the recommendation to postpone was overridden. The green light was given.

At 11:38 a.m. EST, the countdown hit zero. *Challenger* roared to life, its twin, solid rocket boosters lighting up the sky in a blazing trail of fire and thunder. The shuttle rose gracefully, cutting through the blue like a spear of hope. In homes and schools across the nation, eyes were locked on television screens. Children cheered. Teachers smiled. History was in motion.

For seventy-three seconds, it was perfect.

Then, in a moment that would be seared into the collective memory of a generation, *Challenger* disintegrated into a fireball high above the Atlantic.

Seven brave souls vanished in a cruel plume of fire and smoke. In just over a minute, *Challenger* transformed from a symbol of boundless possibility into a heartbreaking reminder of how fragile ambition can be. The explosion tore a hole not just in the sky but in the nation's spirit.

That evening, with the country still reeling, President Ronald Reagan addressed the stunned nation from the Oval Office. His voice was steady, but his words carried the weight of profound sorrow. "They slipped the surly bonds of Earth to touch the face of God," he said, an immortal tribute to those lost.

The tragedy of *Challenger* extended far beyond the immediate heartbreak. It sent shockwaves through the entire aerospace and satellite communications industry. The Space Shuttle program was immediately grounded. All planned launches came to a screeching halt. Satellites, whether commercial, scientific, or government, sat idle on launch pads.

The impact was staggering. NASA, once seen as the unshakable backbone of American space access, had suddenly become a question mark. Satellite operators scrambled to alternate launch providers, many of them foreign, often at exorbitant prices. Insurance premiums for launches skyrocketed. Timeliness that once spanned mere months stretched into years. The dependable cadence of space exploration had been shattered, replaced by uncertainty and delay.

The satellite industry, once flying high on the Shuttle's promise, now found itself rudderless in the wake of disaster. Budgets were strained. Contracts were rewritten. Contingency plans became lifelines. What had been a tightly orchestrated dance of technology devolved into a game of improvisation and risk. And yet, amid the fallout, the industry endured.

Innovation became a necessity. Over time, commercial interest stepped in to fill the void, and a more diversified, competitive market began to take shape. The disaster had dealt a blow, but it also awakened a determination to build safer, smarter, and more adaptable systems for the future.

The Investigation and Reckoning

In the months after the *Challenger* disaster, the nation demanded answers. NASA grounded its shuttle fleet for nearly three years while the Rogers Commission investigated. The cause was clear: mechanical failure, rubber O-rings on the solid rocket boosters had failed in the bitter cold, letting flames ignite catastrophe.

But the deeper failure was human. Engineers' warnings went ignored; red flags were dismissed. The launch didn't fail because the risks were unknown; it failed because they were underestimated. Confidence had hardened into complacency.

Challenger remains one of spaceflight's darkest chapters. Yet from its ashes came lessons forged in pain, engineered into blueprints, etched into memory. Every mission since carries *Challenger's* weight, a reminder in safety checklists, and in the hearts of those who still dare to look skyward.

SPACE SHUTTLE RETURNS

In 1988, the space shuttle program resumed with *Discovery* on September 29, 1988, marking the beginning of a new chapter. *Discovery's* mission was significant: it was the first shuttle flight after the *Challenger* disaster and successfully deployed a communications satellite, *Satcom C-3*, marking the renewed commitment to the shuttle's legacy of satellite deployment and scientific research.

Over the next decade, the Space Shuttle became synonymous with groundbreaking achievements. It was more than just a vehicle for getting astronauts to space; it became the backbone of NASA's scientific and engineering accomplishments. *Discovery, Atlantis,* and *Endeavour* flew dozens of missions during this period: deploying satellites, repairing the Hubble Space Telescope, and conducting experiments that expanded humanity's understanding of the universe. But perhaps the most profound legacy of the shuttle program was its central role in constructing the International Space Station.

The Tragic Loss of the Space Shuttle Columbia

On February 1, 2003, after a triumphant sixteen-day science mission in orbit, Space Shuttle *Columbia* prepared to return home. As it streaked across the morning sky, gliding toward its landing at Kennedy Space Center, millions watched with anticipation, but what was meant to be a routine homecoming turned into a national tragedy—one that would once again shake the foundations of space exploration.

Unknown to the crew and largely dismissed on the ground, a piece of insulating foam had broken off the shuttle's external fuel tank during launch, striking *Columbia's* left wing. It was a moment captured on video but disregarded as inconsequential. No warning systems were triggered. No red flags were raised. The shuttle carried on with its mission, the crew unaware that a fatal wound had already been inflicted.

The small fragment of foam had punched a hole in the heat shield, a breach that would prove catastrophic during the reentry. At 8:44 a.m. EST, just sixteen minutes from touchdown, *Columbia* began to report irregularities. Temperature sensors in the left wing failed. Tire pressure readings vanished. In Mission Control, calm gave way to confusion, and confusion turned to horror.

Moments later, high above the skies of Texas, *Columbia* disintegrated, its fiery debris cutting across the blue sky. All seven astronauts were lost, their final moments vanishing in a storm of flames. The wreckage rained down over hundreds of miles. The nation stood paralyzed in disbelief, hearts shattered, eyes brimming with tears, and unanswered questions echoing into the emptiness of the sky.

Tribute to the Space Shuttle Program

For three decades, the Space Shuttle program stood as a soaring testament to human ingenuity, courage, and the unyielding quest for knowledge. From its inaugural flight in 1981 to its final mission in 2011, the Shuttle was far more than just a vehicle; it became a beacon of our collective dreams, a powerful symbol of what humanity can achieve when fear and doubt are cast aside, and we dare to reach beyond the confines of our world.

Across more than 130 missions, the Shuttle bridged the gap between Earth and the cosmos, carrying us closer to the stars and unlocking mysteries once thought unreachable. It shrank the vastness of space, bringing the universe a little nearer and deepening our profound connection to the infinite beyond.

Though the last Shuttle has now touched down, its legacy continues to blaze across history's sky, etched indelibly into our shared story and lighting the path for future explorers. With every spark of curiosity and every upward glance, a new generation is inspired to carry the torch forward, daring to journey beyond the stars and embrace the limitless frontier that awaits.

CHAPTER 14

REACHING FOR THE STARS

THE BIRTH OF SPACECONNECTION

By the mid-1980s, I had spent years at the center of the satellite indus-try—long days, longer nights, and more lessons than I could count. I'd worked inside three companies that rose to the top of their fields, and each taught me something different: how the business truly operated, where opportunities quietly emerged, and what happened when they slipped through your fingers. The industry was demanding and unfor-giving, but I was fortunate to have a front-row seat to all of it.

I never imagined starting my own company. The idea felt too big, reserved for people with deeper pockets or fewer doubts. But experience has a way of nudging you toward a crossroads. I knew the business, the pressures, and the weight of responsibility. The only question left was whether I had the courage to take the next step.

The timing couldn't have been more intimidating. Hughes, AT&T, and General Electric dominated the landscape—giants with reach, re-sources, and influence that dwarfed anything a newcomer could muster. Challenging them seemed unrealistic, something people whispered about but never dared attempt.

Yet beneath the doubt, a quiet hope persisted. On June 2, 1987, SPACECONNECTION, Inc. was born—not with lavish funding or a large staff, but with determination, a straightforward strategy, and the belief that if we did things the right way, we could earn our place.

Several years earlier, I had enrolled in Pepperdine's MBA program to deepen my expertise in finance, marketing, and strategic decision-making. The training proved invaluable. In the satellite business, every decision carried serious consequences, and a single misstep could put an entire life's savings at risk.

Even so, our first full year surpassed anything I expected. Major clients took a chance on us. International transmissions ran flawlessly. We earned trust the only way that ever works: by doing the job, meeting the standard, and never promising more than we could deliver.

Looking back, SPACECONNECTION succeeded because we stayed committed to our mission. We weren't trying to change the world—just to do our work honestly and carefully. Somehow, that dedication carried us farther than we ever imagined.

Beaming Signals Across the Globe

Our first full year, 1988, began with a global spectacle: *The World in 24 Hours*, a New Year's Eve broadcast that stitched together fireworks celebrations from Sydney to Paris, New York to Rio. What started as a bold concept for NHK, Japan's national broadcaster, quickly became a shared moment across continents—a live tapestry of celebration uniting viewers around the globe.

That momentum carried straight into the Winter Olympics, where unlikely heroes captured the world's imagination. Eddie "The Eagle" Edwards soared into legend despite the odds, and the Jamaican Bobsled Team crashed through both barriers and expectations, winning the hearts of millions. Signals became stories; broadcasts became milestones.

High-stakes diplomacy found its stage, as well. From Israel's fortieth anniversary celebration to a real-time conversation between the Governor of Ohio and Chinese officials, the world's most pivotal moments were carried with precision. The final Reagan-Gorbachev summit marked a turning point in history, and the G7 Summit in To-ronto, along with the election of George H. W. Bush as the forty-first President of the United States, capped a year defined by both political and technological milestones.

On the technological frontier, nothing illustrated the evolution of broadcasting like the 1988 Seoul Olympics. And at Expo '88 in Bris-bane, viewers glimpsed the breathtaking clarity of NHK's revolutionary high-definition television (HDTV), a vision of the future made real through cutting-edge satellite communications.

By year's end, one truth was undeniable: this wasn't just about relaying signals, it was about connecting the world, live and unfiltered. With every flicker on a screen, history unfolded in real time, and we were at its center, quietly shaping how millions experienced it, one signal at a time.

WORLD's FIRST HDTV SATELLITE TRANSMISSION

At the heart of this revolution stood NHK, with a vision that shattered the boundaries of conventional television. In a sleek Tokyo lab, engineers developed a radical analog HDTV system, one that rendered the old NTSC standard almost primitive by comparison. Where Americans had long accepted 525 lines of resolution, NHK delivered a staggering 1,125 lines at thirty frames per second; a visual symphony of detail and motion that transformed the screen into something more: a window into reality.

Creating HDTV was only half the battle. The real challenge was delivering it. Traditional terrestrial broadcasting buckled under the strain of the massive data demands; it was like forcing a waterfall through a garden hose. The solution wasn't on the ground; it was or-biting above. Satellites became the unsung heroes of this technological revolution, delivering the bandwidth and global reach needed to beam breathtaking images across continents and oceans.

Behind the scenes, an international race was heating up. Japan, the United States, and Europe all saw the future on the horizon, and none were content to follow. It was a battle of ideas, of format, and of national pride, with engineers, broadcasters, and dreamers locked in a high-stakes sprint to define the standard of tomorrow.

The defining moment came with the world's first international HDTV broadcast. A historic transmission that bridged continents, connecting New York City to Tokyo, Japan. A signal that defied distance, proving that the future of television had arrived.

Television of the Future

If NHK had sparked the idea of high-definition television, SPACE-CONNECTION (SPC), in collaboration with Hughes Communications, carried that spark across the sky, turning a bold concept into a global reality. The world's first HDTV satellite transmission wasn't just an experiment in a lab; it was a leap into the future, beaming breathtaking images across the Pacific with clarity that stunned even the most seasoned engineers.

The transponders became lifelines of a new era, sending crisp, dazzling images across continents. Where the Japanese Pavilion at Expo '88 had showcased the promise, by June 1, 1989, the promise became reality. For three unforgettable days, New York City flickered to life across Japan in unprecedented detail: the Statue of Liberty sharp against the Manhattan skyline, sunlight glinting off skyscrapers, and clouds drifting over the Hudson with startling realism. Television wasn't just a window anymore; it was a portal.

In Tokyo, passersby stopped in their tracks, mesmerized. In Osaka, department store windows became stages, drawing crowds whose reflections mingled with the glow of this new visual era. Inside the headquarters of Matsushita (now Panasonic), Sony, and Hitachi, engineers and executives watched, knowing they were witnessing history, not just a broadcast, but the unfolding future.

"Japan Begins First Regular Broadcasts of Television's Future." That was the headline in the *New York Times* on Sunday, June 4, marveling at the "cinematic quality" of the images. The report noted how life-like the scenes appeared, so three-dimensional it seemed the Statue of Liberty might reach through the screen and pull viewers straight into New York Harbor.

But this was more than a technological triumph; it was a wake-up call.

Across the Pacific, the reverberations were immediate and profound. In the United States, the birthplace of television, policymakers, broadcasters, and industry leaders scrambled to respond to this new revelation in the broadcasting industry. Congress took notice. America had long led the world in media innovation, but at the dawn of HDTV, it suddenly found itself trailing. Japan had fired the first shot in a new technological race, and the U.S. could no longer afford to stand still.

The implications extended far beyond technology. The first HDTV satellite transmission didn't just raise the standard, it shattered it. Overnight, the old ways of thinking about television were rendered obsolete. A new world of visual storytelling had arrived.

For NHK, SPC, and Hughes Communications, this was more than a milestone. It was the opening chapter of a new era. The impact of that first international HDTV satellite transmission would echo for decades, inspiring a global race to redefine television as the world knew it.

UNITED STATES' FIRST HDTV SATELLITE BROADCAST

Later that year, SPC and Hughes Communications found themselves on the verge of another breakthrough. Fresh off the success of the first international HDTV satellite transmission, the spotlight shifted closer to home: delivering America's first live HDTV broadcast by satellite.

The occasion demanded a spectacle worthy of the moment, and none was more fitting than the final chapter in one of boxing's most celebrated rivalries: Sugar Ray Leonard versus Roberto Durán. On December 7, 1989, under the dazzling lights of the Mirage Hotel and Casino in Las Vegas, history was about to unfold. Inside the ring, two legends prepared for their climactic showdown; outside it, engineers worked tirelessly to bring the action to life with clarity and sharpness never seen before on American television screens.

When the night arrived, fans nationwide tuned in for the legendary fight. But in select venues, viewers experienced something transformative. Prototype HDTV monitors revealed every detail with unprecedented clarity, each drop of sweat, ripple of muscle, and flicker of emotion rendered in astonishing realism. The fuzzy edges and soft focus of standard broadcasts were gone. For those of us watching, the fight felt less like a television program and more like a ringside experience.

Inside the ring, Leonard outmaneuvered Durán with signature speed and tactical mastery, earning a unanimous decision and closing the chapter on a legendary rivalry. Beyond the ropes, another triumph unfolded, one measured not in punches but in pixels. The debut of high-definition broadcasting left audiences spellbound, signaling the dawn of a new era in American television and offering a glimpse of a future where the images on the screen would never look the same.

Turning Vision into Reality

In the aftermath of the groundbreaking 1989 transmission, the impact reached far beyond the Las Vegas boxing ring. As the 1990s dawned, policymakers and industry leaders recognized HDTV as a critical front in the emerging global race for innovation. Japan, led by NHK's ambitious analog HDTV system, had already set an early benchmark. The United States, determined to maintain technological leadership, moved quickly to respond.

Momentum built as Congress, the Federal Communications Commission (FCC), along with key industry stakeholders, embarked on an aggressive campaign to shape the future of American broadcasting. After years of rigorous testing, spirited debate, and a crowded field of proposals, the FCC made a bold decision in 1991 to bypass Japan's analog model and pursue an all-digital HDTV standard. The move permanently reshaped the broadcast landscape.

By 1996, the Telecommunications Act provided the legislative framework for a nationwide transition from analog to digital television. The adoption of the Advanced Television Systems Committee (ATSC) standard enabled breakthroughs such as MPEG-2 compression, allowing high-definition signals to be transmitted efficiently over both satellite and terrestrial systems. Within a few years, major networks like ABC and CBS were broadcasting in stunning 720p and 1080i resolutions, transforming the viewing experience for millions.

At the heart of this transformation was the 1989 HDTV satellite demonstration, a vivid, real-world preview of what high definition could achieve. It captured the attention of engineers, broadcasters, and lawmakers, validating the promise of digital broadcasting and helping catalyze America's commitment to lead into the digital age.

CHAPTER 15

A NEW WORLD ORDER

STAR WARS AND THE COLD WAR DIVIDE

On March 23, 1983, just a year after branding the Soviet Union an evil empire, President Ronald Reagan delivered a nationally televised address that shook the foundations of Cold War strategy. To millions of Americans watching at home, it sounded more like the opening monologue of a sci-fi thriller than a presidential policy announcement.

With calm conviction, Reagan unveiled the Strategic Defense Initiative (SDI), a sweeping plan to create a space-based shield that could detect, intercept, and destroy incoming nuclear missiles before they ever reached American soil. The goal was to render nuclear weapons obsolete.

The media quickly seized on its futuristic imagery, dubbing it "Star Wars." News anchors flashed graphics of satellites firing laser beams from orbit. Political cartoonists had a field day. Critics scoffed and scientists balked, but not everyone was laughing.

The Kremlin Wasn't Amused

For decades, the Cold War had teetered on the knife's edge of mutually assured destruction (MAD)—a terrifying balance where the mere threat of total annihilation kept both superpowers from pushing the button. If either side launched a first strike, both would be reduced to radioactive ash.

But SDI threatened to blow that fragile doctrine to pieces. If the United States could build a functional missile defense system, it might survive a nuclear exchange. That possibility shattered the logic of MAD and sent shockwaves through Moscow.

Behind the Iron Curtain, Soviet leaders feared that Reagan wasn't just bluffing. They worried he was trying to win the Cold War not by firing a shot, but by rendering their arsenal useless. For them, SDI was not a shield; it was a sword cloaked in stars.

The shadow of SDI loomed large when Reagan and Soviet leader Mikhail Gorbachev met for the first time at the 1985 Geneva Summit. Gorbachev, burdened by a faltering economy and eager to reduce tensions, urged Reagan to abandon the initiative. Reagan stood firm. He saw SDI not merely as a defense mechanism, but as a diplomatic lever to protect America or be used as a bargaining chip for peace.

The Cold War reached its boiling point at the Reykjavik Summit in 1986. Across the table, Mikhail Gorbachev made a stunning offer: slash nuclear arsenals to levels once unthinkable. The price? Ronald Reagan had to kill SDI, his "Star Wars" missile shield built on satellites.

Reagan refused. SDI, he argued, was no sword but a shield, and in a shocking twist, he offered to share it with the Soviets. To Gorbachev, it was either naïve or a ruse. He walked.

The summit collapsed. A chance to end the arms race died not over ideology, but a single word: satellites. Reykjavik remains the moment the world nearly stepped back from the brink, only to stumble at the edge.

Melting the Cold War Divide

Though the Reykjavik Summit ended in disappointment, it marked a pivotal moment in history. The tone of the conversation had shifted, moving from threats and missiles to the more hopeful terrain of possibility and reform. By the time of the 1988 Moscow Summit, Reagan and Gorbachev were no longer adversaries separated by a chasm, but partners inching toward common ground.

The Strategic Defense Initiative (SDI), though never fully realized, had served its purpose. It unsettled the Soviets, reinforced American resolve, and subtly tilted the strategic balance just enough to open new diplomatic doors. Beneath the political rhetoric, a new tone was emerging, one marked by cautious trust, a delicate yet powerful shift.

This new trust reached its zenith in one of Reagan's most iconic moments. On June 12, 1987, standing before the Brandenburg Gate in West Berlin, he looked past barbed wire and watchtowers and delivered a challenge that would reverberate through the ages:

"Mr. Gorbachev, tear down this wall!"

Behind the scenes of such history-shaping moments stood Marlin Fitzwater, Reagan's press secretary. From the White House podium, Fitzwater briefed the world through some of the most turbulent hours of modern history. With his calm delivery, sharp command of facts, and steady presence, he became the anchor that guided both press and public through storms of uncertainty.

His skill did not go unnoticed. When George H. W. Bush took office, he asked Fitzwater to remain at his side, making him the only press secretary in U.S. history to be appointed by two presidents. In that role, Fitzwater carried the voice of the presidency through crises and triumphs alike, earning deep respect across party lines.

THE FALL OF THE BERLIN WALL

By the time George H. W. Bush entered the Oval Office, Fitzwater was already a seasoned hand at the helm of history. Together, they would face a world on the cusp of transformation. And before long, that transformation would erupt in one of the most unforgettable nights of the 20th century: the fall of the Berlin Wall.

On November 9, 1989, the world didn't just wait; it watched. For nearly three decades, the Berlin Wall had cast its shadow across Europe, a fortress of stone and wire dividing not just a city, but humanity itself. It stood as the physical embodiment of ideology, fear, and the ever-present threat of conflict.

And then, suddenly, it cracked.

At first, the news trickled out in whispers, as checkpoints opened and border guards stood down. But satellites transformed those whispers into a worldwide shout, impossible to ignore. Within hours, live images flashed across the planet: Berliners swarming the Wall, not with anger, but with disbelief and joy. Families long divided embraced beneath the glare of floodlights, their tears broadcast in real time from the frontlines of history.

Satellites turned the fall of the Wall from a local miracle into a shared human triumph. The chisels striking stone echoed not just in Berlin, but in living rooms from Los Angeles to Tokyo. Fathers lifted children to glimpse what had once been forbidden, and those moments were no longer trapped behind borders; they traveled at the speed of light, captured, transmitted, and replayed for a world that could scarcely believe what it was seeing.

The Wall had fallen, but what echoed louder than the tumbling concrete was the new reality: history no longer belonged to those who wrote the morning papers. Through satellites, history was lived, witnessed, and shared in the instant it was made. On that night, the heartbeat of Berlin became the heartbeat of the world.

THE SPARK THAT LIT THE GULF WAR

The fall of the Iron Curtain promised peace, but history never stays quiet. By the summer of 1990, the Middle East was a tinderbox. Under the desert sun, resentment simmered. Saddam Hussein, scarred by years of war with Iran, accused Kuwait of wrecking his economy by flooding the world with cheap oil.

Words quickly turned to war. Before dawn on August 2, Iraqi

tanks crossed the border, and by daylight, Kuwait City had fallen. Flags changed overnight, and a nation was swallowed whole.

This time, the world didn't learn through delayed reports or second-hand accounts. Satellites made the invasion impossible to hide. Images from space revealed tanks massing on highways. Live feeds showed civilians fleeing, shops stripped bare, and smoke choking the skyline. What once would have taken weeks to confirm was now undeniable in hours.

Hussein's gamble was audacious. For a moment, it seemed he might get away with it. But the same satellites that carried his threats carried proof of his aggression into homes worldwide. Leaders couldn't look away. What began as one nation's gamble became a global standoff, and the message was clear: if Iraq's conquest went unanswered, no border would be safe.

The Call Before the Storm

The phone rang.

It was early January 1991. On the other end, a staffer from NHK Tokyo. His voice was calm, measured, almost too casual for what came next.

"I need a full-time international satellite transponder," he said. "Immediately."

The words jolted me. A request like that wasn't routine; it was monumental. Leasing a full-time transponder wasn't like booking a hotel room. It meant bleeding hundreds of thousands of dollars a month, whether or not a single frame of video ever left the ground. Typically, these deals took weeks of meetings, paperwork, and justification. But this? No explanation. No start date. No project description. Just a cryptic directive dropped like classified orders.

"All right," I said slowly. "What's it for?"

"I don't know," he replied flatly. "My boss wants it secured. That's all I know."

The answer was as unnerving as the request. Vague. Urgent. Unquestionable.

I pressed the point. Did they understand the cost, the risk, the scale of what they were asking? He didn't flinch. "Secure it," was all he repeated. Reluctantly, I processed the order.

And then, nothing.

Days stretched into weeks. The transponder sat in silence, a dark piece of celestial real estate drifting in the geostationary arc 22,236 miles above Earth. Idle. Waiting. Watching. I began to wonder if it would ever be used, or if it was destined to be the most expensive ghost channel in orbit.

Then, one night, the quiet was shattered.

The signal didn't just flicker, it surged. That dormant transponder blazed to life like a Roman candle across the sky.

Screens erupted in an eerie green glow: night-vision footage of tracer rounds ripping across the Baghdad skyline, explosions blooming like cosmic eruptions in the dark, fighter jets screaming overhead. The silence of orbit was gone, replaced by the raw, unfiltered violence of war.

Operation Desert Storm had begun.

In an instant, the mystery was solved. NHK hadn't gambled; they had prepared. Whether by foresight or intelligence, they were ready the moment the first missile left its launch tube.

That phone call hadn't just locked down a satellite channel; it had unlocked a front-row seat to history.

Never before had war been witnessed this way, not in newspapers, not in delayed broadcasts, but live, minute by minute, as if the world were on the battlefield itself.

BROADCASTING FROM THE BATTLEFIELD

In Washington, President George H. W. Bush faced the cameras, jaw clenched, voice unwavering. "This will not stand."

With that declaration, he threw down the gauntlet. Saddam Hussein's invasion of Kuwait would not go unanswered. Behind the scenes, Bush forged something unprecedented: a coalition uniting NATO allies and Arab nations, once divided by centuries of suspicion, now bound together under the banner of justice. Backed by the United Nations, the coalition prepared to strike.

The world held its breath.

At 9:00 p.m. Eastern Standard Time on January 16, 1991, Bush appeared on television screens across America. Operation Desert Storm had begun. A coalition of nations was mobilizing to drive Iraqi forces from Kuwait.

Half a world away, in a high-rise control room in Tokyo, a quieter drama was unfolding. Weeks earlier, an idle satellite transponder had been assigned to NHK, its purpose a mystery. Then, without warning, it sprang to life. Engineers cut to live footage; the world was about to witness war in real time.

At 6:00 a.m. Saudi Arabia time on January 17, the skies over Baghdad erupted.

For the first time, civilians thousands of miles away weren't just hearing about battles; they were watching them unfold. In Tokyo, families huddled around televisions, cafés went silent, and offices paused. Millions of eyes were transfixed as the storm of war lit up their screens. Fighter jets sliced the sky, afterburners glowing like fiery comets. Anti-aircraft fire streaked upward. Oil fields blazed, towers of flame reaching toward the heavens.

The battlefield was no longer distant. Events that once trickled through newspapers and delayed broadcasts now arrived instantly, carried across the globe by satellites. Every moment, every movement, every signal, was transmitted in real time.

For twenty-three days, the Gulf conflict became a live demonstration of satellite power. From Tokyo control rooms to European living rooms, millions watched history unfold in real time and unfiltered.

When the ceasefire was declared on February 28, 1991, the world had changed. The Gulf War was the first conflict of the satellite age. History no longer waited for morning papers or edited footage. The battlefield was here, on screens, in homes, in real time.

THE 1992 PRESIDENTIAL RACE

President George H. W. Bush's popularity soared to almost unprecedented heights after the Gulf War, with his approval ratings rocketing to nearly 90 percent in early 1991. The stunningly swift and decisive victory in Operation Desert Storm, which liberated Kuwait with minimal American casualties, was celebrated as a masterstroke of leadership and military strategy. For a time, Bush seemed almost untouchable; a strong and steady figure commanding respect on the global stage.

Victory abroad, however, couldn't shield him from the growing storm at home. By 1992, America's focus had shifted. The economy had slipped into a deep recession, unemployment was rising, and healthcare costs were spiraling out of control. As the country faced mounting domestic challenges, foreign policy triumphs faded into the background. Meanwhile, Ross Perot, a wealthy independent candidate, further divided the electorate, appealing to fiscal conservatives and moderates disillusioned with both major parties.

Enter Bill Clinton, Governor of Arkansas. His campaign was laser-focused on the issues most pressing to Americans: the economy. His message was simple but powerful: "It's the economy, stupid," a phrase coined by his campaign manager, James Carville.

It struck a chord. On election night, Clinton captured 43 percent of the popular vote and an overwhelming 370 Electoral College votes, putting an end to twelve years of Republican control of the White House.

However, as America chose its next leader, another remarkable story was unfolding beyond its borders.

Democracy at 10,000 Feet

The 1992 U.S. presidential election was a spectacle of democracy, debates, rallies, and votes cast in every corner of the nation. But behind the curtain of American politics, a far quieter, riskier operation was underway. Its purpose: to carry that same election, unfiltered and uncensored, into a place where democracy was little more than a dream.

That assignment fell to **TV Martí**, the U.S. government's Spanish-language broadcaster, tasked with piercing Cuba's state-controlled wall of silence. Their demand was simple but perilous: break through Fidel Castro's iron grip on information and deliver live coverage of America choosing its next leader.

The execution fell to SPACECONNECTION. From production hubs scattered across the U.S., we stitched together a broadcast pipeline with the precision of a military campaign. First, the feeds were moved by satellite. Then, in the Florida Keys, those signals were captured, redirected, and launched skyward, not from a tower, but from a giant tethered aerostat: a massive white helium balloon bristling with broadcast gear, floating 10,000 feet above the ocean.

It was an audacious piece of engineering. High above the Florida Straits, that silent sentinel became the perfect relay, pushing election coverage straight into Cuban living rooms where government jamming was relentless. Each burst of data, each picture frame and word, was a breach in the blockade—a pulse of democracy that no censor could erase.

On that night, Cuban families saw what had been hidden from them: a nation arguing with itself in the open, casting ballots without fear, transferring power in the daylight. It was a forbidden glimpse of democracy in action, carried not by force of arms, but by beams of light and waves of sound.

At 10,000 feet, a balloon became a weapon, one not of war, but of truth.

CHAPTER 16

THE DIGITAL SATELLITE REVOLUTION

THIRD-GENERATION DOMESTIC SATELLITES

Across North America, a new fleet of third-generation satellites took to the skies, ushering in a bold era of innovation that redefined what was technically possible in space-based communications.

RCA Americom led the charge with the launch of *Satcom K1*, equipped with advanced Ku-band transponders tailored for high-speed data transmission. Hughes Communications expanded its Galaxy constellation with *Galaxy V* in 1988, a hybrid C-band and Ku-band powerhouse designed to meet the growing demands of cable and broadcast clients. Just four years later, *Galaxy VI* arrived with even greater transponder strength and reliability, solidifying Hughes's leadership in the sector.

AT&T bolstered its orbital portfolio with *Telstar 302* in 1989 and *Telstar 303* in 1990, both engineered to support corporate communications and a rapidly evolving media landscape, offering improved signal fidelity and coverage. Meanwhile, GTE Spacenet modernized its fleet with *Spacenet II-R* in 1988, followed by *G-Star 3* in 1990, a high-powered Ku-band satellite built for both direct broadcasting and high-volume business data applications.

The SBS series advanced as well, with *SBS-4* and *SBS-5* (launched in 1992) boosting network stability and extending reach to accommodate the skyrocketing demand for satellite television. North of the border, Telesat Canada aggressively entered the U.S. market through cross-border agreements, deploying *Anik E1* and *Anik E2* (both in 1991) to deliver seamless TV and data services across North America.

With the infrastructure now firmly in place, broadcasters stood on the edge of a digital frontier, poised to embrace breakthroughs like MPEG-2 compression, digital multicasting, and eventually, streaming and on-demand content.

The Dawn of the Digital Era

The early 1990s were a hinge in history, a moment when America's flickering screens and scratchy phone lines gave way to something faster, sharper, and infinitely more connected. At the center of this quiet revolution was a familiar player, one long associated with beaming television into living rooms: the satellite. Only now, satellites were about to carry far more than TV shows. They would become the backbone of the digital age.

The breakthrough arrived with third-generation satellites that were smarter, faster, more powerful, and a game-changing ally: MPEG-2 **digital compression.** Where a single analog channel once monopolized an entire transponder, now a dozen digital streams could share the same bandwidth. Images sharpened and sound deepened, proof that geography was no longer a barrier but a challenge already overcome.

This wasn't just about entertainment. It was about information itself. News no longer crawled in on tape or teletype; it arrived live and unfiltered as events unfolded. Corporations stitched their offices together across time zones. Small businesses reached beyond borders. What once looked like science fiction, such as video conferencing, remote collaboration, and even distance learning, was now within reach, hinting at the virtual workplaces and global networks that would define the twenty-first century.

Education was reborn through the sky. Classrooms in forgotten corners of the country received lessons via digital downlinks. Universities broadcast lectures to students who never set foot on campus. Hospitals trained doctors in real time, sending medical expertise across state lines at the speed of light.

By 1993, the media landscape had transformed. Niche networks found their audiences. Sporting events gleamed in crystal-clear definition. Home shopping channels turned televisions into storefronts, driving commerce coast to coast. Satellites were no longer just pipelines for programming; they were launchpads for innovation.

The digital satellite revolution was more than an upgrade in technology. It was a leveling force, collapsing distance, connecting communities, and reshaping how Americans lived, learned, and worked.

PARI-MUTUEL SIMULCASTING OF HORSE RACING

By the late 1980s, horse racing in the United States was on the ropes. Once hailed as the "sport of kings," it was losing ground to a changing world. Attendance was dwindling and cultural tides were shifting. A new wave of distractions, such as cable television, video games, casinos, and state lotteries, pulled the public's attention elsewhere. The message was clear: evolve or become obsolete.

The lifeline arrived in the form of satellite digital compression technology. By the early 1990s, racetracks could beam live races to off-track betting (OTB) sites, casinos, and simulcast venues across the nation. The sport, once tethered to the physical track, now had the power to reach audiences from coast to coast. This was more than a technological upgrade; it was a full-scale reinvention.

Pari-mutuel wagering, where the betting pool determines payouts, was turbocharged by this development. Thanks to real-time, high-quality simulcasts, fans could watch races and place bets from remote locations, experiencing the thrill of the track without ever setting foot on it. The racing experience became more immersive, more immediate, and far more accessible.

Central to this revolution was the adoption of the MPEG-2 standard, a major leap in video and audio compression. It enabled racetracks to transmit multiple high-fidelity feeds efficiently and cost-effectively. Fans could now follow action from multiple tracks, often simultaneously, heightening excitement and driving deeper engagement.

By the mid-1990s, the results were unmistakable. Simulcasting had breathed new life into the sport. Wagering totals soared, the fan base expanded nationally, and horse racing found its digital footing. Regulatory frameworks helped maintain integrity, but it was satellite technology that formed the backbone of this resurgence.

The embrace of satellite digital compression proved to be a defining moment. It preserved the legacy of horse racing during one of its most vulnerable periods, transforming it into a modern, media-savvy enterprise. In a rapidly evolving entertainment landscape, horse racing showed it could still run with the best of them.

A NEW BREED OF SHOPPING TV NETWORK ENTREPRENEURS

In 1982, a new retail revolution was launched when Roy Speer and Lowell "Bud" Paxson co-founded the Home Shopping Network (HSN) in Florida. Their 24/7 live shopping format turned television into an interactive retail platform, blending live demos with charismatic hosts and urgency-driven sales tactics.

QVC entered in 1986 with a polished, customer-focused approach, quickly becoming HSN's main rival. Others followed: Cable Value Network (CVN), later acquired by QVC in 1989; the American Value Network (AVN), which merged into HSN; and Shop at Home (SAH), known for home goods and electronics, whose assets were acquired by Jewelry Television (JTV) in 2006. JTV, founded in 1993 as the American Collectibles Network, evolved into a leading jewelry-focused channel by the early 2000s.

As cable TV expanded, satellite technology became essential. Beginning with HBO's historic *Satcom 1* transmission in the mid-1970s, satellites like *Satcom 3R* (used by HSN), *Satcom 4R* (QVC), and *Galaxy*

III (used by JTV) provided the national reach and reliability needed to distribute live shopping content coast to coast.

Together, these entrepreneurs and technologies transformed TV from a passive medium into a powerful, revenue-generating retail engine, reshaping how America shops, one household at a time.

DIGITAL COMPRESSION FOR CABLE AND DIRECT-TO-HOME (DTH)

The advent of digital compression technology marked a seismic shift in satellite broadcasting. What once required an entire transponder to carry a single analog channel could now support multiple high-quality digital streams. This innovation dramatically expanded programming capacity without the need for additional satellite bandwidth, an absolute game-changer for cable operators and content providers alike. Signal clarity improved, distribution became more efficient, and viewers were treated to a richer, sharper television experience. But the true breakthrough came in the early 1990s with the rise of direct-to-home (DTH) satellite services.

Companies like DirecTV and Dish Network harnessed the power of MPEG-2 compression and next-generation Ku-band satellites to deliver an unprecedented variety of programming directly to consumers. No longer tethered to bulky equipment or complex installations, viewers could now enjoy hundreds of channels through a sleek, eighteen-inch satellite dish installed right at home.

In 1993, Hughes Communications tapped SPACECONNECTION to assist with early technical trials, using a Ku-band transportable uplink. Initially planned as a short-term lease, the project quickly evolved into a pivotal five-month collaboration that helped lay the groundwork for DirecTV's official launch.

When DirecTV went live on June 7, 1994, it introduced America to digital satellite TV with a compact setup, broad content offerings, and crystal-clear picture quality. With that moment, television took a bold step into the digital age, forever changing how viewers accessed and consumed entertainment across the nation.

CHAPTER 17

ORBITAL NIGHTMARES

DREAMS SHATTERED

By the early 1990s, satellite television was rocketing toward a new frontier. Armed with powerful third-generation satellites and the breakthrough potential of MPEG-2 digital compression, broadcasters and cable operators were poised to deliver more channels, sharper images, and lower transmission costs than ever before. The future wasn't just bright, it was blazing.

At the heart of this technological revolution were high-powered satellites built to handle the demands of a rapidly expanding digital world. Leading the charge was *Telstar 402*, a cutting-edge spacecraft engineered by Space Systems/Loral and operated by AT&T's Skynet satellite services. Designed to take over for the aging *Telstar 301*, this new bird was equipped to enhance both C-band and Ku-band services, which were critical frequencies for broadcasters nationwide.

On September 9, 1994, the atmosphere at AT&T's mission control was charged with excitement. Engineers, executives, and technicians crowded around screens, eagerly awaiting lift-off. *Telstar 402* symbolized AT&T's next chapter in broadcasting dominance. Hopes soared as the countdown reached zero. With a deafening roar, the rocket surged skyward. Applause erupted. Years of effort were rising into the heavens . . . until they weren't.

Just minutes into the flight, something went wrong. The rocket's trajectory wavered. Telemetry readings spiked, followed by a blinding flash. An explosion lit up the Atlantic sky. And then . . . nothing. Silence. *Telstar 402* had vanished, obliterated in a fireball above the ocean.

In a single, agonizing moment, the dreams tied to *Telstar 402* were incinerated. The jubilant atmosphere inside mission control turned to stunned stillness. Faces froze. Plans unraveled, sending shockwaves across the broadcast industry. The destruction of *Telstar 402* went beyond technical failure; it was a gut punch to an industry riding high on optimism. Broadcasters who had pre-sold programming and booked satellite capacity suddenly found themselves scrambling, desperate to find alternate transponders to stay on the air. What had begun to feel like a new golden age came to a grinding halt with a chilling reminder: in space, even the best-laid plans can blow up . . . literally.

For AT&T, the fallout was brutal. The financial hit was steep, but the damage to its reputation was even worse. A brand long synonymous with trust, innovation, and rock-solid reliability was now at the center of a catastrophic failure. Customers questioned everything, especially whether AT&T could be counted on in a business where timing, reliability, and precision were paramount.

The loss of *Telstar 402* became a watershed moment. For years, satellites had been the invisible engines powering modern communication, flawlessly delivering TV broadcasts, phone calls, and data around the world. The industry had grown confident, perhaps overly so. The explosion exposed the fragile underbelly of even the most advanced technologies. It reminded everyone from engineers to executives that spaceflight is never routine and progress always carries risk. In the wake of the disaster, satellite providers reexamined procedures, backup plans, and redundancies. The stakes, after all, were measured not just in dollars but in trust.

TELSTAR 402R LAUNCH FROM FRENCH GUIANA

In 1995, SPACECONNECTION received a special invitation from AT&T, being one of their largest customers, to witness a defining moment in satellite history. *Telstar 402R* was poised to rise from the ashes of its ill-fated predecessor. This launch was more than a technical milestone; it carried the weight of reputations, multimillion-dollar contracts, and the relentless drive for reliability in an industry where failure simply wasn't an option.

Rather than a front-row seat at Kennedy Space Center, our journey took us to a far more exotic and unpredictable locale: Kourou, French Guiana, a remote corner of South America more famous for its dense jungles, sweltering tropical fevers, and the notorious Devil's Island than for cutting-edge spaceflight.

The adventure began long before we packed our bags. A battery of vaccines awaited us, more suited to jungle explorers than satellite executives: yellow fever, hepatitis, typhoid, tetanus, rabies. My arm became a dartboard; each injection met with the nurse's cheerful mantra: "Just in case."

Daily anti-malarial pills were handed out like multivitamins, and we were sternly advised to pack insect repellent so potent it could double as "chemical weapons." At this point, all we were missing was a machete and a National Geographic film crew.

We boarded a chartered AT&T flight bound for Cayenne, French Guiana's steamy capital. Stepping off the plane, the air hit as if a warm, wet towel had wrapped around our lungs. From there, we bumped along jungle roads in open jeeps, tree branches brushing our shoulders, the air alive with shrill birdcalls and buzzing wings. It was a tropical paradise: lush, green, and teeming with things that wanted to sting, bite, or crawl under your pant legs and into your socks.

Our destination was the Guiana Space Centre, hidden deep in the jungle like NASA's tropical outpost in the Amazon. Our accommodation was generously referred to as a "hotel." In reality, it was a concrete compound that looked like it could double as a storm shelter. There was no air conditioning, the electricity came and went like a moody teenager, and creaky plumbing wheezed like it needed a medic. Yet somehow, this was touted as the finest resort in Kourou.

At check-in, the receptionist offered a welcome that I remember vividly: "Don't drink the tap water," she warned. It's radioactive, you know, from rocket launch residue." She paused, then added with a shrug that could freeze your blood: "As for tarantulas—they're around. But don't worry. They don't *usually* bite."

That night, I lay under a sagging mosquito net in a room with a cracked ceiling that was pocked with the tracks of giant spiders. The net rustled, not from a breeze, but from something crawling across it. "Tarantulas?" I muttered to myself.

In the gift shop, of course, the best-selling items were preserved spiders, elegantly displayed in glass cases, treasures for the courageous or the borderline insane.

As if the jungle accommodations weren't enough, AT&T arranged a "relaxing" excursion to Devil's Island, the notorious former French penal colony that was still haunted by the ghosts of the damned. If you've seen *Papillon*, you know the kind of hell we were stepping into. If you haven't, imagine Alcatraz baked in tropical heat, swarmed with mosquitoes, wrapped in disease, and surrounded by shark-infested waters; a tropical nightmare masquerading as adventure.

The island was as eerie as it was historic. The ruins, half-swallowed by jungle vines, stood crooked and crumbling, like a jaw of rotted teeth gnawed by time and memory. Moss-covered prison cells yawned open, each one a stone coffin echoing with the ghosts of men who had long been forgotten. Then there was the cemetery, with weather-worn headstones tilting in silence over the graves of those who lost their final battle to madness, disease, and despair.

Lunch, optimistically billed as "gourmet," was served under the full brutality of the midday sun, with not a patch of shade in sight. At the center of the buffet table lay a monstrous, jet-black fish, its gaping mouth bristling with jagged teeth, its oily skin glistening like wet tar. A swarm of flies hovered above it in a buzzing halo, like mourners at a wake. Its cloudy, lifeless eyes stared skyward, as if issuing a silent warning: *You're next.*

By mid-afternoon, the novelty had evaporated. Sunburned, sweat-soaked, and mosquito-bitten, we stood in tense silence for the boat ride back to the mainland. No one said a word, but every glance screamed the same plea: *Get me out of here!*

Back at the hotel, I took a quick shower with my mouth firmly shut. No one needed me glowing in the dark while the launch was underway. That evening, we reconvened at the Guiana Space Center, a gleaming feat of modern engineering carved into the rainforest.

Its location wasn't chosen by chance; perched near the equator, it gave rockets a natural boost, riding the spin of the Earth itself. Every step through the facility hummed with anticipation; this was where science, ambition, and adventure collided in the most spectacular way.

Dusk settled over the jungle, and the crowd held its breath as a voice came over the loudspeaker, slicing through the thick, muggy air with each number of the countdown.

"Trois... deux... un . . . "

Ignition. With a roar that shattered the jungle night, the Ariane rocket surged to life. Flames erupted below as it climbed into the sky, turning the darkness into a canvas of gold and crimson. The earth trembled beneath our feet. *Telstar 402R*, risen from the ashes of its ill-fated predecessor, ascended with majestic force: flawless, determined, and unstoppable.

The crowd erupted. Tension gave way to euphoria. There it was, redemption in motion. Years of planning, heartbreak, and hard-won resilience distilled into a single, perfect ascent.

As the rocket vanished into the heavens, trailing a fiery plume, the jungle seemed to exhale. The mission was a success. *Telstar 402R* was on its way to orbit, a satellite revived, a failure undone, and for us a front-row seat to redemption, etched across the jungle sky. As the flames faded and a pale blue dawn emerged, one thing was certain: AT&T was back in space.

The jungle, the tarantulas, even the looming specter of Devil's Island, all of it slipped away for a moment as we stood in awe of what human determination, science, and second chances can achieve.

By 1996, the U.S. domestic satellite industry was firing on all cylinders. Digital compression unlocked unprecedented channel capacity, unleashing a wave of bold new business models and reshaping the media landscape. Satellite bandwidth was in white-hot demand, fueling everything from nationwide TV networks and a booming cable ecosystem to real-time corporate connections and the pulse-pounding thrill of live sports.

THE DAY THE AIRWAVES DIED

In satellite operations, complacency isn't just dangerous, it's lethal. Far above the sky, beyond every weather report and radar sweep, an invisible predator waits. It has no name, no warning, and no mercy. A rogue flare, a charged particle, a wayward debris fragment, any of it can strike without hesitation. And when it does, systems fail in an instant and years of planning vanish in milliseconds. Space doesn't negotiate. Space doesn't forgive. It punishes swiftly, silently, and absolutely.

January 11, 1997, began like any other winter morning. Cold. Quiet. Unremarkable. Across the United States, coffee brewed in newsrooms, anchors reviewed their scripts, and station engineers ran their usual system checks. At 6:15 a.m. Eastern Time, the nation's broadcasters were preparing to send their first transmissions of the day through *Telstar 401*, a satellite that, until that moment, had never failed them.

Then, everything went black. No countdown, no flicker, and no warning.

High above the Earth, *Telstar 401* had died. It didn't explode. It didn't drift. It simply shut down, as if someone had flipped a switch and cut the cord on one of the most vital lifelines in American broadcasting.

For years, *Telstar 401* had served as a quiet workhorse in the sky. It carried everything from live news and sporting events to national feeds for ABC, CBS, FOX, and PBS. It was invisible to the public, but indispensable to the broadcasters who depended on it. Without it, hundreds of TV stations had no path to their audiences.

A powerful solar storm, launched from the sun's surface, traveled 93 million miles to deliver a fatal blow. It destroyed *Telstar 401's* power core, leaving the satellite disabled. What had been a vital part of the communications network was now just space junk.

In seconds, panic rippled through the broadcasting world. Engineers scrambled to reroute signals. Satellite coordinators barked into phones, trying to reserve the few remaining transponders on neighboring satellites. It was like musical chairs, only this time, the music had already stopped.

The industry had grown fast, agile, bold, and almost invincible in its confidence. The shift from analog to digital had unlocked a revolution: more channels, more bandwidth, and more ambition. But this one event laid bare a sobering truth: for all the genius behind the system, all the millions of dollars spent, it couldn't protect against the powerful forces of the universe.

Inside newsrooms and satellite control centers, people stared at blank monitors and blinking alerts. The feeling was the same everywhere: disbelief, followed by adrenaline, then a cold realization that everything was riding on their next move.

From the outside, America saw no disruption. The morning shows came on as cartoons played for children, but behind the curtain, an invisible war was being waged.

Industry satellite providers rallied together to support their fellow broadcasters in a moment of unprecedented crisis. Companies like PanAmSat, RCA, and SPC collaborated swiftly and selflessly, ensuring critical resources were available to navigate the upheaval.

In an industry defined by competition, rivalries were set aside. The mission was urgent and straightforward: keep America connected. Thanks to this extraordinary coordination, programming resumed. Networks stabilized. The signal endured.

Those behind the scenes would never forget. The sun had fired a warning shot across the bow of modern technology, exposing just how vulnerable even the most advanced systems could be, and yet, in the end, broadcasters stayed on the air, and millions of households remained blissfully unaware of how close they had come to staring at a blank screen, despite *Telstar 401* never returning to service.

CHAPTER 18

THE CROWN JEWELS OF SATELLITE BROADCASTING

THE CORONATION: FROM BLACK-AND-WHITE TO GLOBAL SATELLITE BEAMS

On a June morning in 1953, a young Queen Elizabeth II rode in a gilded coach through the rain-damp streets of London, her coronation marking the first grand ceremony of a modern television age. For most Britons, the event was not experienced from the crowd-lined roads of Westminster but from their living rooms across the country.

Families huddled around small, boxy sets, their screens casting a flickering black-and-white glow. Television ownership was still rare, yet more than 20 million people found a way to watch, gathering in neighbors' homes, local pubs, and village halls. It became Britain's first true shared broadcast experience; a defining moment that not only crowned a monarch but also ushered in a new era of media and national connection.

The BBC carried the pictures, though bound by the limits of early technology and the rigid traditions of royal ceremony. No satellites beamed signals across oceans; instead, film reels were rushed to airports for flights abroad, where foreign audiences might not see the ceremony for hours or even days later. For Britain, however, the coronation marked the moment when the monarchy became inseparable from television, a partnership that would last for Elizabeth's entire reign.

In the United States, live transmission across the Atlantic was still years away. Instead, the coronation was filmed, rapidly developed, and flown overnight, reaching American television screens about sixteen hours later. CBS, NBC, ABC, and the DuMont Television Network* all aired the footage, giving Americans one of their earliest glimpses of a global event on television, though not in real time. The delayed broadcast drew an estimated 85 million viewers across the U.S.; a staggering number that underscored television's growing power to unite people in shared experience.

As the decades passed, so too did the way her reign was seen. By the 1960s and 70s, satellites hovered high above the Earth, turning once-impossible feats into routine broadcasts. No longer confined to reels carried by hand, live images of the Queen's tours, state visits, and Christmas addresses traveled instantly across continents. Viewers in Canada, Australia, India, and Africa could see their monarch at the same moment as those in the United Kingdom. This was the age of the connected world, and Elizabeth, perhaps unknowingly, became one of its first global symbols; her wave, her smile, and her words traveled along invisible satellite highways.

By the time of her Golden Jubilee in 2002, television had undergone further evolution. Digital compression and direct-to-home satellite services ensured that her celebrations reached households with crystal clarity, regardless of their location. Villages in Africa and small towns in North America could tune in as easily as Londoners in Piccadilly. What once required reels of film or massive satellite equipment now happened in an instant, the Queen's image riding beams of light, converted into digital streams that appeared on screens across the world.

*The DuMont Television Network was one of America's earliest broadcasters, operating from 1946 to 1956. Despite its pioneering role, the network closed in 1956 due to financial struggles and a limited number of affiliates, unable to compete with the stronger NBC, CBS, and ABC networks.

When her Platinum Jubilee arrived in 2022, technology had undergone yet another transformation. Drones hovered above royal processions, delivering aerial shots unthinkable in 1953. Ultra-high-definition cameras captured every detail of the pageantry, from the gleam of the crown jewels to the waving flags in the crowd. Streaming services, social media feeds, and smartphones carried the spectacle far beyond the boundaries of traditional television.

And then, later that year, the world watched again, this time in mourning. When Queen Elizabeth II passed away in September 2022, the coverage of her funeral was unlike anything television had ever attempted before. From Westminster Abbey to Windsor, billions across the globe shared in a single, unbroken narrative. Satellite constellations and fiber-optic cables carried not just the sound of solemn hymns or the sight of her coffin draped in the Royal Standard, but the collective silence of a nation and the shared grief of the Commonwealth.

It was a poignant closing chapter to her reign, born into the age of grainy black-and-white broadcasts, concluded in an era when her image and legacy could circle the globe in milliseconds. The Queen had been a constant presence in lives defined by change; her story told not just in history books but through the lens of a camera and the reach of satellites above.

In the end, Elizabeth II's reign was not only the story of a monarch but also the story of television itself, how a crown and a camera grew together, shaping the way a nation, and eventually the world, saw itself.

THE DAY THE WORLD TUNED IN:
THE WEDDING OF CHARLES AND DIANA

On the morning of July 29, 1981, the streets of London were packed with onlookers, flags fluttering in the summer air as a golden carriage made its way toward Saint Paul's Cathedral. Inside, a shy twenty-year-old nursery teacher named Diana Spencer was about to marry the future King of England. But this was no ordinary wedding, and the crowd lining the Mall was only a fraction of the audience.

For the first time in history, a royal wedding wasn't just a national celebration, it was a global broadcast; a live, real-time spectacle transmitted by satellite to every corner of the Earth.

More than 750 million people across seventy-four countries tuned in as the ceremony unfolded. They didn't have to wait for newsreels or next-day headlines. Thanks to the marvel of satellite technology, they were there, watching in real time as Diana stepped from her carriage, her twenty-five-foot train cascading behind her like a wave of silk and history.

It was a moment that marked a turning point, not only for the monarchy but for the media age. Satellite technology, once used primarily for military and scientific communication, had come into its own. This wasn't just a wedding; it was the world's first truly shared television event powered by satellites; a test case for how technology could compress space and time, bringing billions together through a single broadcast.

In Tokyo, commuters paused outside electronics stores to catch a glimpse of the ceremony on glowing screens. In Nairobi, families gathered in community centers powered by generators. In New York, viewers woke early to catch the live feed over breakfast. And in remote villages and urban centers alike, strangers found themselves joined by a single thread; a signal sent into space and bounced back to Earth in milliseconds, carrying the story of a prince, a bride, and the start of something bigger than either of them.

For many, it wasn't just their first royal wedding; it was their first experience of a truly global moment, shared in real time with people they would never meet, in countries they might never visit. It was a glimpse into the future, one where borders no longer dictated who got to witness history, or when.

More than just a fairy tale, the wedding of Charles and Diana became a media milestone: the day the world came together, not in a palace or a church, but in living rooms, public squares, pubs, classrooms, and shops. It was the day satellite television didn't just report history, it became it.

THE DAY THE WORLD STOOD STILL:
PRINCESS DIANA'S FUNERAL

History was made again, though this time in a gloomy and unwelcome way: on the morning of September 6, 1997, black drapes hung over the gates of Kensington Palace. Mourners lined the roads, shoulder to shoulder, many holding white lilies or handwritten notes. The bells of Westminster tolled slowly, solemnly, as a carriage carrying a simple wooden coffin began its journey through the city.

Inside lay Diana, Princess of Wales, a woman who had lived much of her life in front of the world's cameras, and now, even in death, brought the planet to a standstill.

But the crowd in London, vast as it was, represented only a small part of the audience.

An estimated 2.5 billion people around the world watched the funeral live, unified by the invisible reach of satellite technology. In living rooms, cafes, public squares, and rural villages, people from every continent tuned in to witness a farewell not just to a princess, but to a global icon. It was a staggering feat of communication, the culmination of decades of progress in satellite broadcasting.

For nearly four decades, satellites had evolved from Cold War instruments to silent arbiters of global connection. The coronation of Queen Elizabeth II in 1953 was a national event, and the wedding of Charles and Diana in 1981 hinted at the power of live global television. But Diana's funeral was something else entirely. It wasn't just broadcast; it was *shared*, emotionally and instantly, by nearly half the planet.

The coverage itself was a technological triumph, live feeds stitched seamlessly across continents, real-time commentary in dozens of languages, and no delay in transmission as the world said goodbye to the "People's Princess." Satellite technology had reached full maturity, and in Diana's final chapter, it revealed its true potential not as a marvel, but as a medium of global empathy.

What made the moment so powerful wasn't just its reach; it was the feeling that, for a few hours, billions of people were not just watching the same event, but *feeling* the same loss. A princess who had bridged the divide between royalty and humanity in life became, in death, a symbol of our shared experience in the modern world.

Diana's funeral wasn't just a farewell, it was a mirror held up to the planet, reflecting the extraordinary power of technology to connect hearts across borders. It solidified the satellite not merely as a communications tool, but as the backbone of global media, capable of carrying not just information, but emotion.

A ROYAL ODYSSEY: A LIFETIME OF HISTORY

Then, on the day King Charles III was crowned, the world did more than witness a ceremony; it reflected on a lifetime of history carried not just by the weight of a crown, but by the invisible threads of technology that had, for decades, tied the world to the British monarchy.

This was not just a coronation; it was the final act in a global story that had unfolded, chapter by chapter, through the lens of TV broadcast signals.

As King Charles III was crowned, the scene may have looked different, with high-definition broadcasts, social media streams, drone shots, and real-time reactions. But the soul of it remained the same: a ceremony rooted in ancient tradition, transmitted by modern magic. A crown passed not only from monarch to monarch, but from one generation of viewers to the next.

This moment was more than history; it was a living continuation—a tapestry of signal and story, of grief and glory. A royal odyssey, unfolding across decades, had become a mirror of our own lives: joy and heartbreak, triumph and tragedy, all stitched together by the sky above.

CHAPTER 19

SATELLITE NEWS GATHERING

THE BIRTH OF SATELLITE NEWS GATHERING TRANSMISSIONS

Before satellites, live international news was a logistical nightmare. Microwave links, videotape couriers, and costly leased circuits made real-time coverage unreliable, especially in remote locations.

Everything changed in the early 1970s, when satellite technology proved to be a game-changer for the news industry. The first major demonstration came during President Nixon's historic visit to China, a diplomatic milestone the world wanted to witness live.

Hughes Aircraft Company designed a transportable Earth Station, a mobile satellite uplink that could beam live video and audio across the globe. Shipped in modular panels and assembled on-site in Peking, the ten-meter parabolic dish transmitted Nixon's meetings with Chairman Mao and Premier Zhou Enlai with seamless clarity.

The broadcast was a turning point, proving that live, global news was possible. It laid the foundation for modern Satellite News Gathering trucks, forever transforming how the world witnesses history in real time.

KU-BAND: THE USE OF SATELLITE NEWS GATHERING (SNG) TRUCKS

In the early days of satellite transmission, the C-band dominated, ruling supreme in sports and special-event broadcasts. Its massive antennas and reliable signals were the backbone of live television. Yet by the late 1980s, Ku-band emerged as a nimble challenger, offering smaller, more flexible solutions tailored to the evolving needs of news organizations. Operating at higher frequencies, Ku-band made portable satellite dishes and uplink equipment possible, revolutionizing Satellite News Gathering (SNG) by slashing the size and complexity of traditional C-band setups.

Gone were the days of forty-two-foot C-band trailers lumbering down highways. With Ku-band, SNG trucks became compact, nimble, and deployable anywhere, from remote mountain towns to crowded city streets. By the 1990s, these mobile units were the backbone of live reporting, enabling networks to cover breaking news, natural disasters, political events, and more with unprecedented speed and reliability.

The advent of third-generation satellites, featuring higher-powered transponders, further elevated Ku-band's role. Signal quality improved, weather-related disruptions diminished, and SNG trucks became indispensable for live sports coverage. As stadiums, arenas, and coliseums filled with fans, broadcasters demanded real-time feeds from the action. Ku-band-equipped trucks delivered every touchdown, slam dunk, and goal to viewers nationwide, keeping audiences on the edge of their seats.

For years, SNG had been hampered by bulky, transportable uplinks, some of which required fifteen-foot C-band antennas just to deliver broadcast-quality signals. Ku-band dishes, at a more manageable eight feet, offered portability, but broadcasters faced trade-offs: power or flexibility, range or responsiveness. No system had ever bridged the gap until SPACECONNECTION (SPC) introduced Mobile Satellite Connection (MSC), a new enterprise that set out to rewrite the rules.

In 1997, MSC unveiled an ambitious initiative: to design a compact, next-generation uplink that could operate in both C-band and

Ku-band, handle analog as well as digital compression, and pack it all into a sleek, transportable 2.4-meter antenna. The goal was bold: to create a system that was powerful, versatile, and future-ready for the demands of live television.

The result was nothing short of transformative. MSC's dual-band uplink grabbed the attention of major broadcasters, cable networks, and even NASA. In an era when flexibility and reliability were mission-critical, MSC delivered both, dismantling long-standing limitations in a single stroke.

As the industry raced into the digital age, demand for efficient compression and optimized satellite bandwidth skyrocketed. MSC's innovative platform didn't just meet the moment, it defined it. By slashing transmission costs while maximizing bandwidth, it became the go-to solution for networks navigating a fast-changing media landscape. In a short span, MSC earned its reputation as the nation's premier provider of advanced, transportable uplink solutions, setting new benchmarks for performance, efficiency, and innovation.

Then came the call that would push the system to its limits. NASA tapped MSC to retransmit one of the most anticipated moments in modern space exploration: the *Live from Mars* mission. The assignment was straightforward but daunting: retransmit crystal-clear, high-resolution images from the Red Planet to millions of viewers on Earth in real time.

This was no ordinary broadcast. It was a defining moment, a technological marvel in live television, with Mobile Satellite Connection (MSC) as the hub, delivering NASA's *Live from Mars* transmissions from the Jet Propulsion Laboratory in Pasadena to the nation.

BROADCASTING IMAGES FROM MARS

"*Live from Mars* was nothing short of a revolutionary feat, a bold NASA endeavor centered around the historic Mars *Pathfinder* mission, which successfully landed the *Sojourner* rover on the Martian surface on July 4, 1997. For the first time in history, humanity had the opportunity to witness a robotic explorer navigate an alien world in near real-time. This wasn't just a scientific breakthrough; it was a technological spectacle, allowing both scientists and the public to experience the awe-inspiring first movements of a mobile rover on the Red Planet.

The challenge was immense. Transmitting live images and data across the vast, unforgiving expanse of space was an engineering triumph. As soon as *Pathfinder's* lander touched down, it began relaying breathtaking panoramic views of the Martian landscape.

Playing a pivotal role in this historic moment, MSC ensured that these extraordinary images reached audiences nationwide. On July 6 and 9, 1997, MSC successfully retransmitted the trailblazing *Live from Mars* broadcasts from NASA's Jet Propulsion Laboratory (JPL) to the NASA Select Channel, leading science centers, and PBS affiliates, bringing the wonders of the Red Planet into homes and institutions nationwide.

The broadcast marked the first-ever 36 MHz analog transmission using a compact 2.4-meter antenna, a feat previously thought impossible. SPC provided the crucial satellite space segment, utilizing three Ku-band transponders on SBS-6 and a C-band transponder on GE-2, ensuring seamless nationwide distribution of the broadcast.

A Journey Across the Cosmos

Given the staggering distance of over 100 million miles, the signals traveling at the speed of light (186,282 miles per second) took roughly ten minutes to reach Earth. While the images may not technically have been "live" by earthly standards, they represented humanity's closest encounter yet with real-time interplanetary exploration.

For those watching, it felt like science fiction had become reality. Mars, once a distant mystery, was now unfolding before our very eyes, a testament to human ingenuity, the power of satellite technology, and the relentless pursuit of discovery. The pages of history are still being written.

A Bold New Era of Space Exploration

President Trump has stated that America will once again head to the Moon. The renewed ambition marks a cosmic leap in space exploration, building on past achievements while pushing the boundaries of what humanity can accomplish beyond Earth.

With *Artemis I*, a test mission successfully completed in 2022, NASA demonstrated the capabilities of the new Space Launch System, the most powerful rocket ever built, and the *Orion* spacecraft, designed to carry astronauts beyond Earth's orbit. Now, *Artemis II*, slated for next year, will send astronauts around the Moon, paving the way for *Artemis III*, which will land humans on the lunar surface for the first time since 1972.

Unlike Apollo, Artemis is about more than exploration, it's about permanence. NASA, alongside commercial partners like SpaceX, Blue Origin, and Lockheed, is developing technologies that will allow humans not only to visit but live and work on the Moon.

The next logical step is Mars. NASA and SpaceX are actively working on plans for a crewed Mars mission in the 2030s, applying lessons learned from the Moon to this unprecedented interplanetary expedition. If successful, this will be the most significant manned endeavor in history.

America isn't just returning to the Moon, it's advancing towards Mars, the asteroids, and beyond. This new space age will redefine what's possible, proving once again that when America dares to dream big, the universe itself is within reach.

POPE JOHN PAUL II's HISTORIC VISIT TO CUBA

With its groundbreaking compact uplink unit capable of seamlessly switching between C-band and Ku-band frequencies while supporting both digital compression and HDTV transmissions, MSC became the first call made by the major networks and CNN when the world's eyes turned to Pope John Paul II's historic visit to Cuba.

As he traveled from Havana to Camagüey, Santa Clara, and Santiago de Cuba, MSC's dual-band uplink unit stayed one step ahead, ensuring uninterrupted live coverage of each historic moment for viewers across the United States.

For millions watching, the sight of Fidel Castro and Pope John Paul II standing side by side was surreal, as two figures representing vastly different ideologies engaged in a moment of mutual respect. The broadcasts captured the pope's powerful words as he called for "Cuba to open itself to the world, and the world to open itself to Cuba."

It was a lifeline; a bridge between their past and present, between exile and homeland. Some wept as they saw their homeland embrace their faith that had been stifled for decades. Others rejoiced, believing they were witnessing the seeds of transformation being planted.

Stampede on the Road to Santiago

The road to Santiago de Cuba was little more than a winding dirt path, cutting through the Cuban countryside. MSC had been pushing hard, racing against time to set up the next broadcast of Pope John Paul II's historic visit. The air was thick with heat, dust, and the scent of sugarcane; the kind of road where anything could, and did, happen.

Without warning, a stampede of bulls erupted from the brushes, their hooves thundering like distant cannon fire. The lead bull, a beast of muscle and madness, locked eyes with the MSC transportable uplink truck as though it were a challenger in the ring.

The bull slammed headfirst into the truck, the impact shaking the steel frame with a sickening crunch. The front end crumpled, metal screeched, and for a heartbeat, time stood still. Then, as suddenly as they had appeared, the herd disappeared into the distance, leaving only dust, silence, and the wounded vehicle.

Out of the haze, an old farmer came sprinting toward them, his face stricken with pure terror. His eyes darted between the battered truck and the road, as if expecting Cuban officials to materialize and take him away. Before anyone could speak, he dropped to his knees in front of the driver, weeping, begging for forgiveness.

"Please," he gasped in Spanish, his hands trembling. "Take my cattle. All of them. Just . . . just don't tell anyone. They will punish me; my family and I beg you."

The MSC driver, still gripping the wheel, felt his heart tighten. He climbed out of the truck, dusted himself off, and knelt beside the man. With a firm but gentle grip, he pulled him to his feet.

"No," he said, shaking his head. "It's all right. You owe us nothing."

The MSC driver went to work, patching up what they could, checking the uplink equipment, and ensuring the pope's next broadcast would go uninterrupted. The truck bore the scars of the encounter, but it survived. From that day forward, we affectionately called the unit "*El Matador.*"

JOHN GLENN'S RETURN TO SPACE IN HDTV

John Glenn's historic return to space on October 29, 1998, aboard the Space Shuttle *Discovery* was a landmark event in both exploration and broadcasting history. The mission came thirty-six years after his groundbreaking 1962 flight, when he became the first American to orbit the Earth. Now, at seventy-seven years old, Glenn set a new record as the oldest person ever to fly in space.

What made this mission even more extraordinary was how it was broadcast. The launch from Kennedy Space Center in Cape Canaveral, Florida, was one of the first major events to be transmitted live in high-definition television (HDTV), showcasing cutting-edge advancements in broadcast technology.

MSC was entrusted with delivering the HDTV signal, ensuring crystal-clear transmission to key locations, including the National Air and Space Museum in Washington, D.C., NASA Headquarters, the Johnson Space Center, and the Museum of Science in Boston, Massachusetts. Television stations across the country also carried the broadcast, allowing audiences to witness the mission in unprecedented detail.

Glenn's mission reignited global interest in space exploration and proved that human potential is not bound by age. It also underscored the achievement of both the Mercury and Space Shuttle programs, further cementing Glenn's legacy as a true American space pioneer.

Beyond its significance in spaceflight, this event marked a revolution in television broadcasting, proving that HDTV could transform the way the world experiences history. It was a defining milestone, not just for NASA, but for the future of live television, setting a new standard for how outstanding achievements would be seen, remembered, and felt for generations to come.

Setting a New Standard for Live Sports

In the late 1990s, SPC and MSC transformed sports broadcasting, delivering unprecedented clarity and global reach. On August 27, 1999, Monday Night Football became the first NFL game in 720p HDTV on ABC, bringing viewers closer to the action than ever before. Weeks later, the U.S. Open Tennis Finals set another milestone with CBS's first 1080i HDTV broadcast, capturing every serve and rally in stunning detail.

Golf coverage followed suit, as advanced satellite uplinks and 9 MHz digital compression transmitted PGA Tour events to over seventy countries. From The Golf Channel to live tournaments, transportable uplinks ensured flawless, high-definition broadcasts, setting a new standard for live sports worldwide.

Across every major event, satellites and SNG facilities evolved into more than just tools; they became the invisible backbone of a connected world. From stadiums pulsing with the roar of fans to breaking news flashing across continents, these technologies transmitted sights and sounds instantly, connecting humanity around the world.

CHAPTER 20

THE SATELLITE GODS

SATELLITES IN A HOSTILE ENVIRONMENT

It was the Information Age, an era of optimism, innovation, and seemingly limitless possibilities. Yet just beyond that bright horizon loomed a quiet force that few ever considered: the Sun.

Solar storms, those violent eruptions of charged particles, threatened every connection we took for granted. A single flare could scramble signals, fry electronics, or cripple entire networks. In the satellite industry, we never forgot that no matter how advanced our systems became, we were still at the mercy of a star 93 million miles away.

Satellites are marvels of precision. They glide silently overhead, carrying live television, global internet, GPS, weather forecasts, and military communications. Yet for all their sophistication, they remain fragile: a single defect among thousands of components can trigger a blackout, a malfunction, or total loss.

Even more relentless than their intricate design was the Sun, the satellites' most formidable adversary. At the peak of its last eleven-year Solar Maximum in 1989, solar flares erupted with unyielding force. The most powerful, X-class flares, could turn a trusted satellite into drifting wreckage within minutes."

Most people never notice these close calls. The TV stays on, the internet holds steady, GPS keeps guiding. But behind the scenes, engineers scramble in real time, rerouting signals, shifting orbits, rebooting systems, to preserve the seamless connectivity.

Satellites are the quiet backbone of modern life. And yet, every so often, when the Sun flares or a component fails, we're reminded of a sobering truth: our dazzling age of technology still bends to the will of nature.

ROLLING THE DICE IN ORBIT

Buying transponders on a satellite was never just a business deal; it was a gamble, a high-stakes game where the chips were millions of dollars and the casino hovered 22,236 miles above Earth.

The pitch always sounded simple: lease a slice of bandwidth, lock in a contract, and watch the revenue roll in as broadcasters or networks filled the space with programming. But beneath the corporate sheen was a reality every satellite executive understood: no matter how carefully you ran the numbers, you were betting on a machine in space, a fickle market on the ground, and a future no one could predict.

It felt like sliding a lifetime of savings across a Vegas blackjack table. You might hit the jackpot: the transponder could run flawlessly for fifteen years, delivering steady returns. But the odds could turn just as quickly. A solar storm might fry the electronics, or a launch failure could vaporize your investment before it ever reached orbit.

That was the paradox of the satellite business: terrifying and exhilarating in the same breath. You weren't just brokering bandwidth; you were betting on the unknown, gambling in a casino suspended above the Earth.

The stakes were staggering. Because in a single heartbeat, I could be holding the winning hand or helplessly watching it all drift away, as if the heavens themselves had decided to claim all the winnings.

THE SOLAR MAXIMUM THREAT

On Saturday, April 8, 2000, Las Vegas pulsed like the command center of global broadcasting. A record-shattering 115,000 people poured into the National Association of Broadcasters (NAB) Convention. The air buzzed with the certainty that television's future had never been brighter.

But far above the neon lights, a different kind of energy was building—silent, invisible, merciless.

Without warning, the Sun unleashed its fury. A massive solar flare, the most violent in over a decade, erupted, hurling a torrent of charged particles toward Earth at near-light speed. When it struck the planet's magnetic shield, the impact rippled like a sledgehammer against glass. Satellites took the brunt of the blow.

Then came the call no one ever wants to hear. *Telstar 402R*—one of SPC's most trusted workhorses—was gone. One second it hummed steadily, alive with signals; the next, it was dark. Five full-time transponders vanished, including the heartbeat of a twenty-four-hour shopping network reaching 80 million homes.

In an instant, screens across America went black.

And the timing could not have been worse. It was a Saturday afternoon in April, peak sports season. Baseball was in full swing, the NHL and NBA playoffs were in progress, and horse racing feeds jammed nearly every transponder in the sky. There wasn't a single slot of satellite capacity left for our clients.

The hours dragged on. Panic began to seep in. Then—a flicker of hope. By 6:30 p.m. Eastern, the day's marquee broadcasts ended, freeing precious bandwidth. Within minutes, SPC rerouted the shopping network's signal, and several other clients, to alternate satellite capacity. Screens flickered back to life.

When it was over, no one wanted to face that again. We all felt the weight of how close it had come, how easily the situation could have spiraled out of control. A wave of relief washed over us. Hearts still racing, we were simply grateful to have weathered the storm, alive and intact.

High above, Telstar *402R* remained adrift—silent, disabled, a casualty of the Sun's rage. A brutal reminder that in our business, the greatest threats didn't always come from competitors or market forces. Sometimes, they came from the heavens themselves.

LOST IN ORBIT

Morning arrived, and Las Vegas woke with its trademark roar: the hum of slot machines, the rumble of traffic, the pulse of a city that never truly sleeps.

Yesterday's loss of *Telstar 402R's* had landed like a punch to the solar plexus, knocking the wind out of everyone who relied on it. A quarter-billion-dollar satellite, gone in an instant. One moment, it was a guardian in the sky, the next, swallowed by silence. We thought we'd survived the worst. We were wrong.

At 6:45 a.m., the hotel phone shattered the fragile calm. The voice on the line, a *PanAmSat* technician, was tight, urgent, and unmistakably grim.

"We've just lost communication with Galaxy 3R."

This wasn't just any satellite—it was the backbone of the simulcast empire, the one bird you could not afford to lose. Not during peak season. Not when every fleeting second of live horse racing funneled billions into sportsbooks, from Vegas to off-track betting parlors across the nation.

The timing was merciless. It came on the heels of devastation—still reeling from the total wipeout of every SPC transponder on *Telstar 402R*—but this news was even worse.

SPC had just lost five full-time transponders on Galaxy 3R, the heartbeat of its biggest client. These were no ordinary channels—they carried live horse racing to every wagering hall and betting screen in America. Cut them, and the simulcast empire didn't falter—it flatlined.

Then came the second call. My client, the architect of the simulcast operation, spoke with unnerving calm, but the tension seeped through every word.

"All the casino bosses are calling," he said. *"They're asking if they should stop taking bets."*

He sought clarity. I had nothing to offer. Millions were vanishing—gone, second by second—and I was powerless to stop it.

The phone rang again. *PanAmSat*, controlled and tense.

"We are in emergency mode—no telemetry, no command, we think it's tumbling."

"Tumbling." The word hit like a wrecking ball. The satellite was in a full free-spin, cartwheeling through space with no control, no guidance—just chaos.

Backup? None. Spare transponders? Gone. *Telstar 402R* was already a ghost, drifting silently through the cosmos. Finding five transponders on the same satellite? Impossible.

Then the phone rang again. My client's voice was taut, sharp.

"They're pressing for a decision. The bosses . . . do they keep taking bets?"

I didn't hesitate. *"Tell them to hold."*

But Vegas doesn't pause. Every second costs money, every heartbeat erodes trust. And the truth: unless we pulled off a miracle, the entire simulcast empire was barreling toward complete devastation.

THE EDGE OF OBLIVION

This was no longer just a technical failure. It was a fuse burning toward financial disaster, a crisis unfolding at the very heart of the gaming universe.

Post time was 10:00 a.m. Pacific, 1:00 p.m. Eastern. Less than two hours remained to pull off the impossible.

Then the order came down from the casino bosses. Sharp. Final. Irrevocable.

"Shut it down. Stop taking bets."

In an instant, Las Vegas's heartbeat stopped. The city that never sleeps descended into a waking nightmare.

The back-room kingpins barked into phones, voices fraying with urgency. Investors prowled dim corridors, demanding answers no one could give. Racing lounges, once alive with cheers and anticipation, froze. It was as if someone had yanked the plug on reality itself. Screens went black, betting boards fell silent, and a billion-dollar industry ground to a halt.

Across America, wagers stalled as panic gripped the venues. Terminals locked. Odds disappeared. Some stared in disbelief, others slammed fists in frustration, shouting into the void.

The first post approached, ninety minutes away. Sports lounges that usually rang with laughter and the rhythmic chatter of ticket printers now echoed emptiness. Big screens flickered, then surrendered to darkness. Waitresses paused mid-step. Floor managers whispered behind clenched jaws, powerless.

Upstairs, behind locked doors, blinking red lights ablaze, panic crystallized. Investors stared at spreadsheets hemorrhaging zeroes. Phones lit up like Christmas trees. Financial dread swept through the room like an icy wind. The system's spine had snapped.

For the satellite company's owner, this wasn't just lost revenue—it was my whole identity vanishing. Losing our biggest client, the cornerstone of our business, reputation, and purpose, wasn't a setback; it was annihilation.

By 9:00 a.m. Pacific, the last traces of hope were gone. No signal, no explanation, and no answers. Only the hollow echo of loneliness. This wasn't just about the company anymore. It was about my life.

Within twenty-four hours, the destruction defied belief. Losing a single transponder on a satellite was devastating. Losing five full-time transponders was catastrophic. Losing ten was an annihilation. Struck by the Sun's fury, SPACECONNECTION was thrown into helpless chaos, reeling from two crippled satellites and staring down a $90 million commitment of transponder leases.

At 9:36 a.m. Pacific, 12:36p.m. Eastern, it was twenty-four minutes to post time.

The hotel phone rang, slicing through the silence. My heart sank. I had nothing left to lose—except everything. A voice came through instead, steady, strained with relief.

"We've regained control. Galaxy 3R is back in service. Full telemetry. All transponders operational."

At 10:00 a.m. Pacific, 1:00 p.m. Eastern, the gates sprang open. Ticket printers roared. Cheers erupted, not because fortunes had been won, but because the heartbeat of racing had returned.

For 171 agonizing minutes, the satellite gods had held their breath. Then, like a thunderbolt from the heavens, *Galaxy 3R* roared back to life, resurrected from the brink. The lifeblood of an entire industry surged once more. From the shaded paddocks of Saratoga to the steel gates of Santa Anita, horses shifted into place. Betting halls from Atlantic City to Reno erupted in relief. Screens flickered on. Silence gave way to celebration.

After gut-wrenching hours of silence, the tension finally shattered. First *Galaxy 3R,* then, later that day, *Telstar 402R*—two satellites, teetering on the brink of oblivion—roared back to life. The satellite gods had intervened. The lifeblood of an entire industry surged once more, pumping energy and relief through every wagering hall, every broadcast screen, every heartbeat of the empire.

During the NAB Convention weekend of April 8 and 9, not everyone hit the jackpot at the slots, but in the satellite world, we all walked away winners. Las Vegas may be built on chance and the house's edge, but in those final moments, we witnessed something far rarer than luck.

We bore witness to a miracle—the answer to our prayers.

We didn't just win. We endured. We defied failure.

In a city where the house always wins, the satellite gods turned the tables on fate itself. They went all-in and dealt the single hand that mattered most: **SURVIVAL!**

CHAPTER 21

THE DIGITAL AGE

THE SATELLITE HIGHWAY TO THE INTERNET

At the turn of the millennium, two revolutions converged: the satellite revolution and the internet revolution. Together, they ignited a communications explosion that would reshape the way humanity connected. Sleek satellites spun silently above the Earth, their circuits humming with promise, while personal computers from visionaries like Apple and Microsoft brought the digital frontier to desks and living rooms worldwide. The stage was set for a new era, one in which ideas leapt across borders in an instant, friendships blossomed across continents, and the planet began to pulse as a single, interconnected community.

The internet was no longer a laboratory experiment or a niche curiosity; it was a sprawling, living network of possibility. Yet the Earth itself was a barrier. Oceans, mountains, and deserts stood between people and the information they needed to access. Fiber-optic cables could only stretch so far, and laying new lines across continents, or under turbulent seas, was slow, costly, and often impossible.

Above it all, satellites waited in geostationary orbit, tens of thousands of miles from the surface. These were no ordinary machines; they were invisible lifelines, capable of carrying voices, data, and images across the globe in the blink of an eye. In those delicate circuits and orbiting transponders lay the promise of a world without boundaries.

By 2000, the fusion of satellites and the internet was creating a world that was faster, smaller, and more connected than anyone had imagined. We weren't just sending emails or transferring files; we were building the invisible highways of the Information Age, with satellites forming the backbone of the digital revolution that would define the twenty-first century.

A DECADE OF CONSOLIDATION AND TRANSFORMATION

The early 2000s brought a seismic shift to the domestic satellite communications industry. What had once been a fast-moving, entrepreneurial arena—built on the ingenuity of independent resellers, uplink providers, and transponder brokers—was reshaped almost overnight by a wave of mergers, acquisitions, and global consolidation. Globalization, digital innovation, and mounting economic pressures collided, and suddenly survival required scale, capital, and direct control over satellite assets.

Amidst this turbulence, SPACECONNECTION saw what many others did not. While competitors clung to old business models or hoped the storm would pass, SPC recognized that the era of privately-owned, independent resellers was coming to an end. The company had always thrived on reading the industry's tides, anticipating shifts in technology, understanding customer needs before they were spoken, and navigating the high-stakes world of satellite inventory with a steady hand. Now, it became clear that the next chapter demanded alignment with a major satellite operator.

The broader industry confirmed that instinct. Major U.S. operators were rapidly consolidating. PanAmSat, Loral Skynet, and GE Americom were absorbed by larger global players. SES entered the U.S. market in 2001 by acquiring GE Americom, creating SES Americom. PanAmSat—anchored by its powerful Galaxy fleet—was purchased by KKR in 2004 and later folded into Intelsat in 2006, merging enormous transponder inventories under one global giant. Intelsat also acquired

Loral's satellite assets in 2004, rebranding them as Intelsat Americas and tightening the landscape even further.

In that pivotal moment, SPC recognized the timing was right. As consolidation accelerated and the industry transformed around them, their actions were guided not by fear, but by careful foresight. SPC knew that to continue delivering the reliability, service, and broadcast excellence their customers relied on, they would benefit from the backing and stability of a larger, established operator.

Telesat Canada reached the same conclusion from the opposite direction. Seeking to expand its video distribution reach and establish a strong presence in the U.S. market, Telesat acquired SPACECONNECTION in 2004. It was a strategic fit on both sides—Telesat gained a respected team, a loyal customer base, and decades of industry trust, while SPC secured its future in an industry that had fundamentally changed.

MARKET FRAGMENTATION

For decades, cable had ruled the living room, with satellite TV as its steadfast ally, beaming premium programming into homes across the nation. But beneath the polished surface, cracks were forming. The industry was hurtling toward a disruption that would leave network executives scrambling, satellite operators reeling, and audiences forever transformed.

It began quietly, with small internet platforms offering downloadable content, dismissed in corporate boardrooms. "Who would watch a movie on a computer screen?" executives scoffed. They didn't see the tide rising: a generation of tech-savvy, impatient consumers, hungry for on-demand entertainment, methodically rewriting the rules of television.

Then streaming erupted like a wildfire. Viewers were no longer tethered to rigid schedules. Services like Netflix unleashed the binge-watching era, allowing audiences to devour entire seasons at their own pace. Younger viewers, in particular, abandoned traditional TV in droves, triggering a steep decline in viewership and, with it, advertising revenue, the lifeblood of broadcasters.

By the mid-2000s, streaming was on the brink of upending the entire television ecosystem. Cable and satellite providers, once untouchable giants, suddenly found themselves scrambling to respond. Direct-to-home (DTH) satellite services, the revolutionary force of the previous decade, struggled to integrate streaming options fast enough to keep pace.

Audiences were no longer passive recipients of network-curated schedules; they had become the architects of their own entertainment. Traditional television's rigid structures crumbled, replaced by a world where consumers dictated what to watch, when to watch, and on which device. Control had shifted, irreversibly, into the hands of the viewer, and there was no turning back.

The Streaming Showdown

Streaming services like Netflix, Hulu, and Amazon Prime Video changed the game, offering on-demand content at a fraction of the cost of cable, all without commercials. Early streaming in 2007 was clunky, limited, and plagued by buffering. But as broadband improved, so did streaming's appeal. The ability to binge-watch entire seasons of television shows shattered the rigidity of traditional TV schedules.

Cord-cutting surged, shaking the industry's old guard. Satellite providers like DirecTV and Dish Network, once dominant, faced a losing battle. Streaming required no contracts, no satellite dishes, and no costly hardware, just an internet connection. Desperate to stay relevant, DTH providers scrambled to bundle streaming into their packages, hoping to stem the tide of departing customers, but viewers had tasted freedom.

Major networks launched their own services—HBO Max, Disney+, Peacock, and others—each vying for subscribers with exclusive content. Licensing deals became strategic weapons in the fierce race for digital dominance. Budgets soared as streaming platforms poured billions into original films, series, and even live events. The last stronghold of traditional TV, live sports, shifted online, with Amazon acquiring Thursday Night Football rights and leagues exploring direct-to-consumer streaming, bypassing broadcasters entirely.

THE RISE OF STREAMING (2010-2020)

By 2010, satellite broadcasting had already reshaped the global media landscape, providing reliable, high-quality transmission for TV networks, cable programmers, and direct broadcast satellite (DBS) services, such as DirecTV and Dish Network. However, the rise of broadband internet connections, streaming platforms, and evolving consumer habits sparked a dramatic shift in how content was distributed and consumed.

At the start of the decade, satellite TV providers and traditional broadcasters still held a commanding presence. Networks like ABC, CBS, NBC, and FOX relied on satellite feeds to distribute content to their affiliates. Cable programmers such as ESPN, HBO, and Discovery Channel also depended on satellite for national distribution to cable headends and DBS services.

However, streaming platforms like Netflix, Hulu, and Amazon Prime Video were gaining traction, driven by improvements in broadband and growing consumer demand for on-demand content. For the first time, satellite-based TV distribution faced a legitimate competitor, one that didn't require a dish or cable box. During this time, broadband speeds were improving, but satellite still provided unmatched reach, particularly in rural areas where high-speed internet was unavailable.

By 2015, cord-cutting was in full swing. Consumers were dropping expensive pay TV packages in favor of a la carte streaming options. Traditional DBS services, such as DirecTV and Dish Network, began losing subscribers at an alarming rate. Despite these adaptations, satellite broadcasting's role in consumer TV distribution began shrinking, as viewers increasingly shifted to over-the-top (OTT) services: streaming platforms that deliver content directly to consumers over the internet, bypassing traditional cable or satellite TV providers.

The early 2020s marked the beginning of a new era in television—the Streaming Wars. Media giants unveiled their own platforms, battling for viewers who now had an endless array of choices at their fingertips. Satellite TV saw its influence fade, but satellites still played a role in delivering content to the world.

THE FUTURE OF SATELLITE BROADCAST DISTRIBUTION

The satellites were still there, silently orbiting thousands of miles above, carrying the pulse of America's live sports, breaking news, and special events to millions of homes. ABC, CBS, NBC, and FOX all depended on them. But down on the ground, the air was charged with quiet upheaval.

Cable programmers were pushing compression to its breaking point. With high-efficiency video coding (HEVC), they could pack more HD, and now 4K, content into the same bandwidth. It was elegant, efficient, and necessary because viewers were no longer patient.

Direct-to-home services like DirecTV and Dish still held ground in rural areas, but the streets of digital entertainment were crowded. Streaming had arrived, relentless and irresistible. Entire libraries on demand, watched on phones, tablets, and TVs with no schedules, no cable box, and no waiting. DTH providers scrambled to bundle streaming into their packages, but viewers had tasted freedom, and once that door opened, it couldn't be closed.

Even the old stalwarts felt the tremor. Shopping channels like QVC and HSN, once untouchable fixtures of satellite lineups, now saw younger audiences drift online. To stay relevant, they adapted to live streaming, interactive apps, and e-commerce integration. Loyalty alone wasn't enough.

And in pari-mutuel simulcasting, the game itself was undergoing a transformation. Horse racing was no longer bound to the track. Fans wagered from phones, tablets, and living rooms across the country, following every stride and jump in real time. The industry poured into digital experiences, mobile platforms, and interactive features, chasing a future that had already arrived.

In this fractured, fast-moving landscape, the satellites remained a backbone, invisible but indispensable. Yet the real story was the scramble on Earth, broadcasters, programmers, and networks racing to keep pace with an audience that had already left the old rules behind.

CHAPTER 22

LIFELINES FROM SPACE

THE STRATEGIC ROLE OF GEOSTATIONARY SATELLITES

From the moment the first geostationary (GEO) satellites took their place in orbit, the flow of information on Earth was transformed. They became invisible partners to television networks, carrying live sports, breaking news, and primetime dramas into homes from coast to coast. Entire cable empires—CNN, ESPN, Discovery, and HGTV—rose on their shoulders, their signals beamed to head ends across the country before spilling into millions of living rooms. Later, providers like DirecTV and Dish Network bypassed the ground entirely, streaming entertainment directly from orbit to small rooftop dishes.

But the strategic role of GEO satellites extends far beyond television. In times of disaster, when earthquakes rattle cities, hurricanes shred coastlines, or when cyberattacks cripple terrestrial networks, it is often satellites that endure. Where fiber lines snap and cell towers collapse, GEO links continue to hum with life, giving first responders the secure connections they need to coordinate rescues, deploy resources, and broadcast urgent alerts. In such moments, a clear, unbroken satellite signal can mean the difference between chaos and survival.

Modern GEOs are far more capable than their predecessors. Outfitted with powerful transponders, steerable spot beams, and advanced

Ku- and Ka-band payloads, they now deliver broadband internet, extend 5G coverage, support distance learning, and connect communities unreachable by fiber. Thanks to breakthroughs in compression, such as HEVC, they can transmit more data at higher quality while conserving spectrum, making them both efficient and cost-effective.

A NEW ERA: SATELLITES IN MOTION

Above us, satellites race silently across the sky—unseen yet indispensable, reshaping how the world stays connected. Low Earth orbit (LEO) constellations streak just a few hundred miles above Earth, blazing fast enough to deliver fiber-like speeds and latency as low as twenty to forty milliseconds, enough for cloud computing, video calls, and real-time gaming. Thousands of satellites, from *Starlink* to *OneWeb*, *Project Kuiper*, and *Telesat Lightspeed*, reach corners of the world that fiber and cellular networks can't touch.

Between LEO and GEO, medium earth orbit (MEO) satellites fill the gap, offering broad coverage, low latency, and resilience. Perfect for enterprise networks, mobile backhaul, and remote connectivity, MEO strengthens the orbital web that binds the planet.

But modern connectivity isn't about LEO, MEO, or GEO alone; it's their harmony that makes the magic happen. GEO holds steady as the backbone of broadcasting, MEO delivers reach and reliability, and LEO streaks overhead with low-latency agility. Together, they weave an unbroken, always-on network, orbiting tirelessly above, keeping the world connected in real time.

When disaster strikes, satellites become lifelines. Hurricanes flatten cities, earthquakes sever fiber, wildfires erase roads and power lines, but signals from space endure. They guide rescue teams, deliver real-time imagery, and keep vital information flowing. In the chaos of catastrophe, when everything on the ground fails, the sky remains vigilant, silent, and unwavering.

HOW SATELLITES KEPT AMERICA CONNECTED ON 9/11

Smoke choked the streets of Lower Manhattan. The Twin Towers were gone, reduced to twisted steel and dust, and with them, the city's lifeline of communication. Phones were dead. Radios were silent. Chaos stretched in every direction.

High above Florida, Air Force One roared through the sky. President George W. Bush gripped the armrest, eyes scanning briefings, listening to advisors. On the ground, uncertainty reigned: Was Washington next? Were more attacks imminent? Terrestrial networks had failed, but one lifeline remained. Secure satellite links carried his words to the White House, to NORAD, and to military commanders. Invisible threads from orbit held the nation's command together as the world below teetered on the edge.

In New York, first responders wove through smoke and rubble. Every step was perilous, every moment, critical. Radios had failed, but portable satellite phones pierced the silence, carrying orders, reports, and a glimmer of hope. Lives depended on those signals, on the unseen eyes and ears above.

Television crews scrambled, trucks jostling through debris-filled streets. Fiber lines were gone. Microwave towers had been destroyed. Satellites became their conduit, sending live images of ground zero to a stunned world, shaping how history would remember the day.

Even the financial heartbeat of the nation relied on satellites, keeping banks, trading floors, and government agencies alive when terrestrial networks had vanished.

Amid falling steel and collapsing voices, satellites endured. Unseen and unyielding, they stitched the broken world together. In America's darkest hours, they were more than technology; they were the nation's lifeline.

ISRAEL-HAMAS CONFLICT

At dawn on October 7, 2023, Israel awoke to horror. Hamas fighters struck across the border in a coordinated assault, launching rockets, breaching defenses, and leaving over 1,200 Israelis dead, with dozens taken hostage. A nation long confident in its security found itself suddenly shattered.

Israel struck back with overwhelming force. Airstrikes tore through Gaza, followed by ground operations aimed at dismantling tunnels, weapon caches, and command centers. Civilians on both sides faced devastation and displacement as violence spread like wildfire.

But beyond the smoke and chaos, another battle raged in the skies above. Israel's survival depended not just on troops and aircraft, but on an invisible shield of satellites and advanced defense systems. As rockets, missiles, and drones rained from Gaza and beyond, from Lebanon, Yemen, and even Iran, Israel's multi-layered defenses came alive.

The Iron Dome became the world's symbol of interception, catching short-range rockets in midair flashes. Beyond it, David's Sling and the Arrow systems reach higher and farther, neutralizing ballistic missiles before they can strike. Satellites feed these systems, providing eyes, ears, and instant communication, turning chaos into coordinated action.

This space-powered shield traces its roots to the 1980s, when Reagan's Strategic Defense Initiative envisioned satellites tracking threats from orbit. Israel turned that vision into reality. Every rocket stopped, every missile neutralized, every drone shot down carries the quiet signature of satellites overhead.

In modern warfare, where threats strike fast and from every direction, satellites give Israel its most precious asset: time. Time to see, time to react, and most importantly, time to survive.

ORBITS OF HOPE: WHEN NATURAL DISASTERS STRIKE

Disasters strike without warning. In late September 2024, Hurricane Helene barreled across the Atlantic, a 150-mph juggernaut tearing through coastal towns and inland communities in the Southeastern United States. Cell towers snapped, fiber-optic lines were submerged, and emergency responders struggled to coordinate rescues. Entire communities were cut off from the world just when communication mattered most.

Far above, a silent fleet of satellites held vigil. SpaceX's Starlink constellation, orbiting in low Earth orbit, became a lifeline. Compact, self-aligning terminals no larger than a pizza box allowed residents and first responders to connect instantly, powered by generators, solar panels, or even a car battery. Within hours, over 2,000 terminals were deployed via trucks, boats, and drones, restoring connectivity in some of the most severely affected areas.

One rescue exemplified their impact: a woman in labor and a group of stranded storm victims were trapped on an isolated island, cut off by the hurricane's surge. Starlink-enabled drones pinpointed their location, and an evacuation boat guided by real-time satellite data reached them within hours, bringing the woman to safety. For Hurricane Helene, satellites were the quiet heroes keeping the wheels of rescue, relief, and survival turning.

Just months later, in January 2025, Los Angeles County faced another devastating emergency, this time from fire. Wildfires tore through hills and neighborhoods, destroying over 2,000 structures and displacing 100,000 residents. Smoke, heat, and destruction took down terrestrial communication networks, leaving many without access to information or emergency services.

Once again, satellites answered the call. SpaceX provided Starlink terminals to reconnect isolated communities, enabling residents to check on their loved ones and emergency responders to coordinate efforts in real time. A partnership with T-Mobile activated direct-to-cell services, restoring SMS and 911 texting capabilities even in areas without working towers. For the residents and first responders navigating smoke-choked streets, satellites weren't just technology; they were lifelines.

From hurricanes to wildfires, from storm-swept islands to smoke-choked cities, satellites remain the silent guardians above, unseen, unwavering, and ever watchful, keeping communities connected, rescue teams informed, and hope alive when disaster strikes.

BREAKING NEWS: TRUMP HAS BEEN SHOT

The fairgrounds in Butler, Pennsylvania, was electric. Thousands of voices rose in anticipation as former President Donald J. Trump stepped onto the stage. Then, a sharp crack shattered the roar. Gunshots.

Secret Service agents dove, tackling him to the ground. Screams shredded the air. The crowd scattered, tumbling over one another, panic rippling outward like shockwaves. Every heartbeat stretched into an eternity.

Above, satellites streaked silently across the sky, indifferent to the chaos below. They scanned, tracked, and transmitted with unerring precision. Every smartphone buzzed, every camera flared alive, every news feed ignited across the globe. By the time the crowd even realized what had happened, the attack had already been captured, streamed, and broadcast worldwide.

A bystander's hands shook as their phone streamed live. Seconds later, millions were watching. Newsrooms exploded into action. Directors barked into earpieces: "Go live! Go live!" Trucks roared to life. Ku-band uplinks fired. Screens flickered with multiple angles, each feed verified in a blur of controlled chaos. Reporters sprinted to their cameras, adrenaline coursing through their veins, voices cutting through the panic and confusion.

And above it all, satellites held their silent vigil. Invisible yet indispensable, they connected networks, bridged continents, and ensured the story reached every corner of the globe—instantaneously, unfiltered, and unstoppable. In that moment, the world's most critical technology proved itself not just helpful, but essential, turning catastrophe into a story that no one would miss.

Breaking News: A Tale of Two Timelines

To grasp how profoundly news has transformed, consider another moment in history: the day President Lincoln was shot, when the world waited anxiously as news slowly crept across continents.

In 1865, news of President Abraham Lincoln's assassination spread slowly, across continents. The transatlantic telegraph cable, temporarily out of service, left Europe in the dark. It took twelve days for a steamship to carry the tragic message overseas. From there, it trickled through telegraphs and newspapers, arriving in fragments, with each update carrying the weight of uncertainty.

Fast forward to 2024. When former President Donald Trump was shot during a rally in Butler, Pennsylvania, there was no waiting, no uncertainty. Within seconds, a bystander's smartphone captured the event and streamed it live on Facebook. Satellites overhead relayed the footage instantly to newsrooms, screens, and devices around the globe. A story that once took twelve days to reach Europe now reached millions of eyes and ears in less time than it takes to read this sentence.

Today, satellites transmit signals across oceans, smartphones transform ordinary citizens into frontline reporters, and the internet delivers breaking news, emergency alerts, and live events to anyone, anywhere, in real time. What once trickled painstakingly across continents now floods the world in a relentless torrent of instant information, a striking testament to the astonishing evolution of communication.

CHAPTER 23

WHAT LIES AHEAD

BROADCASTING IN THE DIGITAL AGE

The world of television is in the throes of a historic transformation. Once defined by scheduled programming and a handful of channels, the medium is now being reshaped by the rise of over-the-top (OTT) streaming services. Viewers increasingly demand content on their own terms, favoring internet-delivered, on-demand programming over traditional linear broadcasts. Networks are racing to keep pace, investing heavily in their own digital platforms, while technology giants such as Google, Apple, and Amazon enter the arena with original content and sophisticated distribution systems that rival even the most established broadcasters.

Cable television, too, has been swept up in this digital wave. The era of cord-cutting, in which consumers abandon long-standing subscriptions, has forced providers to reinvent themselves. Many now offer hybrid models that bundle high-speed internet with access to popular streaming services. These approaches combine the reliability of traditional TV delivery with the flexibility and convenience of on-demand content, appealing to audiences who want the best of both worlds.

Direct-to-home (DTH) satellite services are also adapting. Once seen as the pinnacle of consumer choice, providers now integrate OTT streaming into their satellite offerings, allowing subscribers to access both live television and internet content through a single, unified platform. Smart set-top boxes have become central to this transition, supporting streaming apps alongside conventional channels. Some operators are experimenting with Internet Protocol Television (IPTV), replacing traditional satellite signals with broadband delivery, opening the door to cloud DVR, interactive applications, and customized recommendations that enhance the viewing experience.

By 2025, satellite broadcasting occupies a more specialized yet indispensable role. Networks still rely on satellites for live coverage of global events like the Olympics, the Super Bowl, and the World Cup. For direct broadcast satellite (DBS) providers such as DirecTV and Dish Network, satellites continue to serve rural households and regions with limited broadband access. Meanwhile, internet-based services like Netflix no longer depend on satellites for delivery, but emerging low-Earth orbit constellations promise to expand broadband access worldwide, indirectly fueling the streaming revolution.

In today's digital age, broadcasting has evolved from a one-way transmission model to a vibrant ecosystem defined by choice, personalization, and interactivity. Traditional television has not vanished; it has adapted, reinventing itself to coexist with the growing dominance of internet-delivered entertainment. What remains constant is broadcasting's essential role: connecting audiences to culture, sports, news, and stories. Only now, those connections traverse more platforms, reach more devices, and offer more possibilities than ever before.

THE CURRENT STATE OF HDTV AND BEYOND

HD television took flight in 1989, a bold leap that transformed how the world watched. What started as a flicker of clarity became a global standard, but HDTV was never the finish line. Streaming, 4K, and now 8K have turned screens into portals, pulling viewers deeper into every story.

1080i and 720p broadcasts once dazzled; today, true 1080p on Netflix, YouTube, and Amazon Prime outshines them, proving that streaming has overtaken traditional TV in clarity and consistency. 4K brought four times the detail, making sports, films, and live events shockingly real. 8K, sixteen times sharper than 1080p, hints at near-photorealism, with Japan's NHK showing the way during the 2020 Olympics. AI upscaling and smarter codecs are turning what once seemed impossible into everyday viewing.

Behind it all, satellites hum silently in orbit, the unsung engines of this evolution. Smarter compression, IP-based delivery, and high-throughput networks enable the transmission of massive streams with precision and reliability. They bridge oceans and continents, survive disasters, and deliver moments as they happen.

Television is no longer just watched, it's experienced. And above it all, satellites power the magic, turning every broadcast into an immersive, unbroken thread that binds a connected world together.

SATELLITES AND THE FUTURE OF WARFARE

Not long ago, the Gulf War unfolded live in living rooms around the world, transmitted by satellites. Operation Desert Storm wasn't just a conflict; it was a revelation. Night-vision missiles streaked across the sky, armored divisions moved like chess pieces, and embedded journalists brought the frontlines home. Satellites had transformed warfare, not just how it was fought, but how it was seen.

Today, satellites are more than observers; they are the battlefield itself. GPS-guided strikes, encrypted communications, real-time surveillance, and logistics all rely on geostationary and low earth orbit constellations. In Ukraine, Starlink terminals became lifelines when terrestrial networks failed, connecting soldiers and civilians alike. The true high ground is now above us, invisible yet decisive.

The future promises even more dramatic change: commanders viewing 16K ultra-high-definition imagery, drones and robots guided with millisecond precision, and holographic war rooms projecting battlefields commanders can walk through virtually.

Yet satellites also hold the promise of prevention. Their unmatched surveillance, rapid-response capabilities, and global transparency could serve as tools of deterrence, diplomacy, and disaster relief. The more the world can see, the harder it becomes to act without consequence. In this paradox, satellites may not only help fight wars; but also prevent them.

The New Golden Dome

In January 2025, President Donald Trump signed an executive order launching a bold new chapter in U.S. defense strategy: the "Golden Dome." Inspired by Ronald Reagan's Strategic Defense Initiative of the 1980s and Israel's Iron Dome, this program is designed to protect the homeland against next-generation threats, from hypersonic weapons to long-range ballistic missiles launched by adversaries like China, Russia, and Iran.

Unlike traditional defenses, the Golden Dome pushes the battlefield skyward. Hundreds, potentially thousands, of satellites, outfitted with ultra-sensitive sensors, would orbit the Earth as sentinels, detecting missile launches and instantly relaying targeting data to interceptors capable of neutralizing threats before they reach U.S. airspace. Complementing this orbital network, roughly 200 armed platforms would create a 24/7, multi-layered shield spanning the globe.

Where Israel's Iron Dome excels against short-range rockets, the Golden Dome envisions a shield against intercontinental and hypersonic threats traveling at speeds exceeding Mach 5. It represents a paradigm shift: space itself as the ultimate high ground, where battles are waged at the speed of light.

Critics once ridiculed Reagan's SDI as "Star Wars," yet it laid the foundation for breakthroughs in missile tracking, radar, and satellite communications—advances that now fuel the Golden Dome. This next-generation system is a technological marvel and a symbol of renewed collaboration between government, private aerospace companies, and defense innovators.

The Golden Dome is more than a defense system; it is a statement: tomorrow's threats won't wait, and the battles of the future will be waged at the speed of light. In the Space Age, satellites are no longer silent observers; they are guardians, enablers, and sometimes the first line of defense.

A GLIMPSE INTO THE SATELLITE-POWERED LIFE OF TOMORROW

If someone had told you decades ago that tiny spacecraft orbiting hundreds of miles above Earth would one day become your invisible roommates, you might've laughed. Today, you probably rely on them before your feet even touch the floor. Welcome to the future, quietly powered by satellites.

Imagine waking up in a connected world where satellites above are already working for you.

As dawn breaks, you rise from a restful sleep. But it wasn't just rest, you were monitored, optimized, and gently recharged by a smart bed that adjusted itself in real time. While you slept, biometric data traveled seamlessly through space via a constellation of low-Earth orbit satellites to your personal healthcare platform. Before you're even out of bed, your day is already calibrated.

Your mirror doesn't just reflect your face; it reflects your condition. It greets you with personalized insights: sleep scores, hydration levels, and early health alerts. All of it streamed from space. Your lights brighten gradually, synced to your internal clock. Downstairs, your coffee is already brewing, perfectly timed to your sleep cycle.

Unseen, but always working, a silent network keeps everything in sync. Your home hums with satellite-connected intelligence. The fridge reorders what you're running low on. The shower starts at just the right temperature. Your security system stays online, even in a blackout, always alert, always connected.

When it's time to head out, your autonomous car is already waiting. No need for traffic updates, satellites feed it live weather, road, and construction data. It adjusts your route, speed, and even your playlist based on signals from your wearable health monitor.

At work, global collaboration feels local. Colleagues from Dubai, Berlin, and Buenos Aires appear instantly on your screen, with no buffering or lag. That is the power of bonded satellite broadband. Critical files, medical scans, and sensitive data are all sent securely through encrypted, space-based channels, delivered at the speed of light.

After sunset, satellites keep working while you wind down. Your lights dim to a soft glow. Your oven preheats on cue. Your air system shifts to match outdoor pollen levels. Even if a storm takes out local infrastructure, your home stays online. Emergency alerts, live security feeds, backup power are all guided by orbiting technology.

So, the next time your groceries arrive just as you walk through the door, or your doctor calls before you even feel sick, remember, it's not magic, but a quiet little helper, circling far above, making sure everything here on Earth works just right.

CHAPTER 24

SATELLITE COMMUNICATIONS AND ITS LEGACY

THE FUTURE OF SATELLITE TRANSMISSIONS

We are entering a new space age, defined not by rocket launches but by the silent streams of data flowing from above. Satellites are rapidly becoming the unseen infrastructure of everyday life, linking people, powering industries, and opening new frontiers. In the years ahead, satellite transmissions will deliver more than signals; they will provide solutions. From bridging the digital divide and guiding autonomous vehicles to enabling humanity's return to the Moon and Mars, these orbiting systems are poised to shape the very trajectory of our future.

The once-clear boundary between satellite broadcasting and internet streaming is rapidly dissolving. Hybrid networks are emerging that combine satellite coverage with terrestrial fiber, 5G and 6G systems, and low-earth orbit (LEO) constellations like *Starlink* and Amazon's *Project Kuiper*. Satellites will act as high-speed data relays, delivering reliable connectivity to areas that remain off the grid today. Artificial intelligence is revolutionizing media delivery, enabling satellites to transmit content tailored to each viewer. Algorithms will analyze preferences, habits, and real-time behavior to create personalized, on-demand experiences at unprecedented speed.

Ultra-high-definition and immersive media will push the limits of imagination. While 4K has become commonplace, tomorrow promises 8K, 16K, and even holographic broadcasts, with satellites handling the massive data loads these formats require. As space exploration accelerates, reliable communication across cosmic distances will become essential. Next-generation satellite networks will enable real-time contact with spacecraft en route to the Moon, Mars, and beyond. Relay satellites orbiting planets and moons will form the backbone of an interplanetary internet, ensuring astronauts and mission control remain connected across the solar system.

Future space colonies will rely on satellites not only for communication but also for navigation, weather forecasting, surveillance, and even power distribution. These orbiting systems will serve as invisible lifelines, sustaining interplanetary civilization and mirroring Earth's digital infrastructure. Tomorrow's satellites will think independently: AI-powered constellations will make real-time decisions, adjust orbits, detect anomalies, and manage entire networks without human intervention, transforming them from tools into autonomous agents of progress.

The evolution of satellite transmissions is more than a technological leap; it is the bedrock of a global renaissance in how we live, connect, and explore. From streaming personalized content at lightning speed to defending our skies, powering autonomous systems, and reaching across galaxies, satellites have evolved far beyond mere signal carriers. They are the silent architects of a bold new age, igniting a celestial renaissance where science fiction becomes engineered reality, and the stars shift from distant dreams to destinations within reach.

A LEGACY IN ORBIT: CELEBRATING EIGHTY YEARS OF SATELLITE INNOVATION

For centuries, distance was an insurmountable barrier. Messages traveled at the speed of a horseback rider, ships took weeks to cross the ocean, and entire civilizations remained isolated, unaware of events unfolding beyond their borders. Even as technology progressed, radio and television signals remained tethered to terrestrial limits, restricted by towers and cables that struggled to reach the most remote corners of the globe.

As we reflect on the legacy of satellite communications, 2025 marks three pivotal anniversaries, each representing a breakthrough that redefined how we connect, communicate, and experience the world:

Eighty Years Since Arthur C. Clarke's Vision of Geostationary Satellites

In 1945, science fiction writer and futurist Arthur C. Clarke proposed an idea that would change the course of history: geostationary satellites, spacecraft positioned 22,236 miles above Earth that could provide continuous global communication coverage.

At the time, it was dismissed as ambitious fantasy. Yet, Clarke's vision became reality, forming the backbone of modern telecommunications and enabling everything from live television to global internet access. His concept not only revolutionized communication but also inspired engineers and space pioneers to turn fiction into reality.

Sixty Years Since *Early Bird*: The First Commercial Communications Satellite

On April 6, 1965, *Intelsat I*—famously known as *Early Bird*—soared into orbit as the first commercial communications satellite. Launched by NASA for the Communications Satellite Corporation (COMSAT), *Early Bird* delivered the first-ever live transatlantic television broadcast, bridging the continents between Europe and the United States.

For the first time, major global events could be shared across oceans in real time, shrinking the world and laying the foundation for today's interconnected society. This moment marked the dawn of instant television, voice, and data communication, proving that no distance was too vast to overcome.

Fifty Years Since the First Live Multi-Carrier Televised Event via U.S. Domestic Satellite

On August 9, 1975, the United States ushered in a new era of broadcasting with the first-ever live multi-carrier televised event via domestic satellite. Utilizing Western Union's *Westar I* satellite, KXAS-TV in Dallas/Fort Worth delivered a historic broadcast to American viewers, showcasing the untapped potential of satellite transmission.

This was more than a breakthrough. It was a revolution unfolding in real time. Satellites carried sports, news, and entertainment across the nation with flawless precision, forging a superhighway of signals that powered nonstop sports, round-the-clock cable, and twenty-four-hour news. Americans didn't just watch; they experienced history live as it happened.

THE NEXT FRONTIER

For those of us who grew up in the Space Age—who watched Sputnik usher in a new era, held our breath as humanity first set foot on the Moon, and marveled as satellites transformed the way we see the world—we have witnessed one of history's most extraordinary chapters unfold before our eyes.

What began with grainy analog transmissions has evolved into high-definition digital broadcasts, and now into ultra-high-definition 16K, virtual reality, and even holographic experiences. Satellites remain the lifeline of global communication, enabling emergency response, defense operations, remote education, and deep-space exploration.

With space technology advancing at an unprecedented pace, the next breakthrough in satellite communications is just over the horizon. Advanced constellations are weaving a digital web across the sky, powering real-time disaster response networks that save lives and space-based defense systems that ensure global security. We are stepping into a future once reserved for science fiction.

The past eighty years of satellite innovation guides us toward a world where every person, every place, and every moment is connected.

No matter how remote, distance will no longer divide us when satellites unite us.

And when the next defining moment arrives—whether it's humanity's first step on Mars or a breakthrough we can't yet imagine—the world will be watching… **LIVE VIA SATELLITE.**

APPENDIX

ROBERT M. PATTERSON
BIOGRAPHY

Robert M. Patterson is a visionary entrepreneur whose pioneering work helped shape the modern satellite communications and broadcasting industries. Over a distinguished five-decade career, he transformed bold ideas into groundbreaking innovations that forever changed how the world experiences live sports, breaking news, and global events.

His journey began in 1969 with a subsidiary of Hughes Aircraft Company—the builder of the satellites that delivered the Apollo 11 Moon landing to a global audience. Inspired by that historic broadcast, Patterson committed his career to advancing satellite technology—not merely following its evolution; but leading it.

In 1975, Patterson orchestrated the first-ever live, multi-carrier television broadcast via domestic satellite in the United States—a landmark achievement that laid the foundation for modern live broadcasting. That same year, in addition to Major League Baseball, he arranged the first satellite transmissions of regular-season NHL and NBA games; and coordinated the first daily satellite news distribution service for independent television stations nationwide.

Years ahead of the digital revolution, Patterson demonstrated in 1978 that two digital signals could be transmitted simultaneously over a single satellite channel—a pioneering breakthrough unveiled to Alaska's state legislators, offering an early glimpse of the future in digital distribution.

Throughout his career, Patterson played a central role in delivering some of the most-watched events in television history. He arranged the satellite distribution of the landmark *Frost/Nixon* interviews, the most-viewed TV political interview series ever aired; coordinated coverage of the Holmes vs. Cooney heavyweight title fight, one of the era's largest pay-per-view events; and secured global transmission rights for the 1984 Los Angeles Summer Olympics, enabling 24/7 coverage watched by over a billion people.

As founder of The SPACECONNECTION, Inc., Patterson built one of the industry's premier satellite transmission company; serving broadcasters, corporations, government agencies, educational institutions, and entertainment networks worldwide. Among his many innovations, he facilitated the world's first international satellite transmission of high-definition television (HDTV), followed by the first domestic HDTV broadcast in the U.S.—milestones that established a new standard of sharpness and clarity in television broadcasting.

Patterson also played a crucial role in transmitting major global events, including the Reagan-Gorbachev summits, the fall of the Berlin Wall, and Operation Desert Storm, as well as key NASA missions such as Space Shuttle flights, *Live from Mars,* and John Glenn's return to space. In association with SPACECONNECTION's affiliated company, Mobile Satellite Connection, he spearheaded the development of the first transportable uplink units capable of transmitting both analog and digitally compressed SD and HDTV signals via compact 2.4-meter antennas in both C-Band and Ku-Band, redefining mobile television versatility.

In 2004, Patterson sold SPACECONNECTION to Telesat, Canada's leading satellite operator.

A native of Queens, New York, Patterson earned a B.A. from Franklin Pierce University, where he later served on the board of trustees.

In 2002, the university honored him by naming its television facility the Patterson Broadcasting Studios. He also holds an M.B.A. from Pepperdine University, where he was recognized as "Top Entrepreneur" by his graduating class.

TECHNICAL INFORMATION

1) Explanation for S-Band: S-band is often used for long-distance space communications, including transmissions to and from spacecraft millions of miles away, low-Earth orbit (LEO) satellites, and spacecraft telemetry, tracking, and command (TT&C). S-band operates in a lower frequency range (2-4 GHz), which experiences less attenuation as the signal travels through space.

2) Explanation for Terrestrial Facilities - From the late 1940s until the 1970s, American Telephone and Telegraph Company (AT&T) used coaxial cables, landlines, and microwave relay facilities to transmit television and radio signals across the country. However, microwave transmissions operate in the line-of-sight mode, meaning the transmitter and receiver antennas must have an unobstructed path between them. As distances increase, the curvature of the Earth becomes a significant obstacle, limiting the span that the signal can travel. Consequently, microwave signal distances could only travel about thirty miles, depending on geographical barriers.

3) Explanation for Geostationary Positioning: Communications satellites are strategically positioned in a geostationary orbit, where they appear virtually fixed relative to the Earth's surface. Maintaining this position involves a delicate balance between the centrifugal force, generated by its velocity of approximately 7,000 mph, to conform with the Earth's rotation and its gravitational pull. Precise orbital maneuvers counteract minor gravitational perturbations from celestial bodies like the Sun and Moon to prevent the satellite from drifting outside its intended figure-eight pattern. Small gas jets, powered on board fuel, are periodically fired by the thruster to maintain the satellite's proper orbital position.

4) Explanation for the Satellite Delay: The signal had to travel a staggering 44,472 miles (i.e., 22,236 miles to the satellite and the same distance back to Earth). At the speed of light, this journey takes about 270 milliseconds or roughly a quarter of a second. In contrast, the landline signal only had to travel about 1,000 miles at the speed of light between Milwaukee and Dallas, making the transmission virtually instantaneous. However, the average viewer at home would never have noticed the time delay, as they weren't watching both feeds simultaneously.

5) Explanation for the Simultaneous Transmission and Recovery of Alternating Pictures (STRAP): Two complete TV programs are carried over a single transponder by time-sharing the video-alternating pictures, then rebuilding the two continuous programs at the receive end. Think of it as analog time-division multiplexing for television. The audio for both programs rode separate FM subcarriers, so each program's audio remained continuous, even while its video "alternated turns" between every frame (or field).

6) Explanation for C-band Frequency: This is one of the oldest and most widely used frequency bands in satellite communications, and it has been the backbone of international and domestic satellite broadcasting for decades. Frequency ranges for satellite use: uplink (Earth-to-satellite): 5.925-6.425 GHz and Downlink (satellite-to-Earth): 3.7-4.2 GHz.

Explanation for Ku-band Frequency: This is a portion of the microwave frequency spectrum widely used for satellite communications, primarily broadcast television, very small aperture terminal (VSAT) networks and direct-to-home (DTH) services. Frequency ranges for satellite use: uplink (Earth-to-satellite): 14.0-14.5GHz and downlink (satellite-to Earth): 10.7 - 12.75 GHz.

7) Explanation for Frequency Coordination: The FCC requires frequency coordination for transportable uplink facilities to prevent interference with local terrestrial carriers. This is particularly important because C-band frequencies are also used for terrestrial microwave services. To address this, specialized companies conduct coordination studies using state-of-the-art software, comprehensive databases, and detailed interference analyses. Their role is to ensure coexistence between satellite uplink facilities and incumbent point-to-point microwave systems, safeguarding operations in the 6 GHz band without harmful interference.

8) Explanation for the Telecommunications Act of 1996 and ATSC: In the early 1990s, U.S. broadcasters were transitioning from analog to digital technology. To facilitate this shift, the FCC created the Advanced Television Systems Committee (ATSC) to standardize digital TV in the United States. With multiple digital TV formats competing globally, U.S. policymakers sought a clear standard to ensure widespread adoption, ultimately mandating the ATSC standard for digital transmission.

The ATSC standard defined how digital TV signals would be encoded, transmitted, and decoded, incorporating key features such as MPEG-2 digital compression, which allowed multiple channels to share a single frequency band; HDTV support with 720p and 1080i formats and enhanced audio formats, closed captioning, and interactive capabilities.

The combination of HDTV technology, the ATSC standard, and the Telecommunications Act of 1996 provided the framework for the nationwide transition from analog to digital broadcasting, culminating in the official end of analog broadcasts in 2009.

9) Explanation for MPEG-2 Digital Compression: MPEG-2 (Moving Picture Experts Group - Phase 2) is a digital video compression standard introduced in the mid-1990s. It was designed to reduce the amount of data needed to transmit or store video and audio while maintaining high picture and sound quality. MPEG-2 became the backbone of digital television (DTV), satellite broadcasting, and cable TV. MPEG-2 analyzes movement between frames and sends only the information about what has changed, not the entire picture; thus, it shrinks massive video files into manageable streams without sacrificing too much quality.

10) Explanation of Solar Flares: Solar flares are sudden, intense bursts of radiation and energy that erupt from the Sun's atmosphere. The Sun follows an approximately eleven-year cycle, with the peak of activity known as Solar Maximum. During Solar Maximum, the Sun's magnetic fields are at their most active, and solar storms occur far more frequently, sometimes multiple times per day.

These storms can disrupt signals between satellites and ground stations, interfering with communications and data transmissions. In severe cases, solar storms can cause irreversible damage to satellite systems, potentially leading to complete system failure. The flares themselves emit electromagnetic radiation, which travels at the speed of light, about 186,000 miles per second.

11) Explanation of Lost Communication and Spinning Out of Control: When a satellite loses communications with its ground control center, operators can no longer send commands or receive telemetry data. This means they can't monitor the satellite's status or position, or communicate with it. If communications cannot be established and the spinning can't be controlled, the satellite may become unusable, leading to system failure. A spinning satellite may drift from its designated orbit (i.e., the figure 8 box), increasing the risk of collision with other satellites, or it could cause interference with adjacent satellites. The chances of re-establishing communications with a spinning out-of-control satellite are somewhere between slim and none.

12) Explanation for Low Earth Orbit (LEO) Satellites: Low earth orbit (LEO) satellites orbit between roughly 100 and 1,200 miles above the Earth, completing an orbit in about 90–120 minutes. Their proximity allows for very low latency—ideal for real-time communications, broadband, and Earth observation—but their small coverage area requires large constellations to provide continuous global service.

Medium earth orbit (MEO) satellites, orbiting between 1,200 and 22,236 miles, cover larger regions with fewer satellites and have moderate latency, making them well-suited for navigation systems like GPS and regional broadband networks. While LEO prioritizes speed and low-latency connectivity, MEO balances coverage and operational longevity, providing complementary roles in modern satellite communications.